Pelican Books
CONTINENTAL DRIF

D1151441

D. H. Tarling is Senior Research Officer in the
Department of Geophysics and Planetary Physics at
the University of Newcastle upon Tyne. His work
is mainly research on palaeomagnetism as applied to
continental drift, but he also lectures on geophysics
and continental drift. He has worked in the Middle
East, North Africa, Australia, North America, the
Pacific islands, New Zealand, and other parts of the
world, mainly in connection with the study of
continental drift. He is a Fellow of the Geological
Society of London, the Royal Astronomical Society,
and the Geological Society of America, and a member
of the American Geophysical Union.

M. P. Tarling, who is not a scientist, worked with her
husband on the manuscript of *Continental Drift* to
present the information in a readily comprehensible
form. She has worked on a wide variety of journals
from *Modern Poultrykeeping* to *Modern Medicine*
and accompanies her husband on most of his
palaeomagnetic expeditions, on rare occasions even
wielding a sledgehammer in the interests of the
subject.

CONTINENTAL DRIFT

A Study of the Earth's Moving Surface

D. H. Tarling and M. P. Tarling

Penguin Books

Penguin Books Ltd, Harmondsworth,
Middlesex, England
Penguin Books Australia Ltd, Ringwood,
Victoria, Australia
Penguin Books Canada Ltd,
41 Steelcase Road West,
Markham, Ontario, Canada

First published by Bell 1971
Published in Pelican Books 1972
Reprinted 1974
Copyright © G. Bell & Sons Ltd, 1971

Made and printed in Great Britain by
Richard Clay (The Chaucer Press) Ltd
Bungay, Suffolk
Set in Linotype Baskerville

CONTENTS

LIST OF FIGURES

LIST OF PLATES

PREFACE

This book is written for anyone interested in the Earth, to provide an introduction to, and summary of, the subject for the student scientist and an overall picture for the general reader with no scientific background. We have not simplified the arguments or the scientific observations, but have tried to present them in an easily comprehensible manner without technical jargon necessitating constant reference to a glossary. In such a brief historical review we have mentioned only those who, we consider, have made the most significant contributions to the development of the idea of continental drift, and for any one example of evidence used in this book there are many tens or hundreds of examples not mentioned. It is, in fact, this range of subjects and ideas which makes the study of continental drift of such continuing and absorbing interest.

We wish to thank Dr H. I. S. Thirlaway and J. Young of U.K.A.E.A., Blacknest, for permission to reproduce fig. 40(a). The following figures are redrawn from published diagrams by kind permission: Fig. 3 from du Toit: *Our Wandering Continents*, published by Oliver & Boyd; Fig. 4 from Wegener: *The Origin of Continents and Oceans*, published by Methuen and Co. Ltd; Fig. 8(a) from their Royal Society paper *The Fit of the Continents around the Atlantic*, by Sir Edward Bullard, J. E. Everett, and A. Gilbert Smith; Fig. 8(b) from a contribution by S. Warren Carey to the Tasmanian *Continental Drift*

Symposium. We are also indebted to Dr H. Lister of the University of Newcastle upon Tyne for reading the manuscript.

We hope that the reader of this book will experience some of the excitement of the most recent discoveries about the world on which we live.

Newcastle upon Tyne   DON AND MAUREEN TARLING
July 1970

Did the continents once fit together like a gigantic jig-saw? The answer to this puzzle is now causing a revolu-tion in the geological sciences comparable with the effect of Darwin on the biological sciences a century ago. Our increased knowledge of the surface geology of the world, combined with the recent great advances in the investiga-tion of the interior of the Earth, make it possible to say that not only must the continents have once fitted to-gether and have since drifted apart, but the solid floors of the oceans themselves take part in this movement.

The idea of continental drift is, in fact, very old and the correspondence between the coastlines of Africa and of South America appears to have been known to most of the explorers of the Age of Discovery. In *Novanum Organum*, published in 1620, Sir Francis Bacon commented that such similarities could hardly be accidental, but he did not offer any interpretation of the observation. Shortly after-wards, in 1658, the Frenchman, R. P. François Placet, wrote a memoir (*La corruption du grand et petit monde, ou il est montré que devant le déluge, l'Amerique n'était point separée des autres parties du monde*) in which he suggested that the Old and New Worlds became separated following the Flood. Another theological scholar, Theodor Lilienthal (1756), also found biblical evidence for the separation of a single landmass into the present continents. In 1800, Alexander von Humboldt, the pioneer explorer of Mexico and the United States, still retained the idea that the Atlantic was essentially a huge river valley whose sides had been separated by the great volume of water over which Noah's Ark had sailed.

The first observation of the geological, as opposed to the geometrical, similarities of the continents on either side of the Atlantic was made by Antonio Snider-Pelligrini, in 1858. He described in his work, *La Creation et Ses Mystères Dévoiles*, the fitting together of the continents bordering the Atlantic, which J. H. Pepper used in 1861 to explain the occurrence of identical fossil plants in the coal deposits of both Europe and North America. In Snider's work, he gave the first diagram showing the continents bordering the Atlantic before and after their separation (Figure 1).

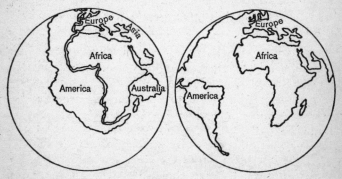

1. Snider's reconstruction of 1858

This was the first diagram of the fit made to explain the similarities of 300-million-year-old fossils in the coal deposits of Europe and North America.

The nineteenth century, of course, saw many fruitful ideas emerging on a vast range of subjects including continental drift. Charles Darwin, on his epic-making voyages, found convincing signs of vertical movements of land masses, but did not see evidence for large-scale hori-

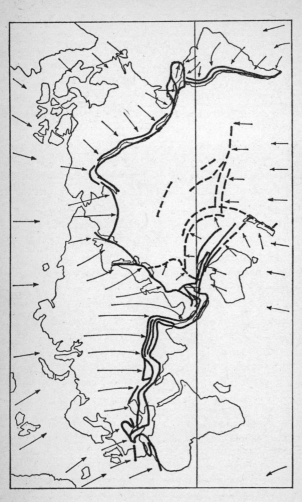

2. *The direction of continental drifting according to Frank Taylor*

In 1910, Taylor suggested that the continents must be moving in the direction of the arrows in order to crumple the rocks into our modern mountains (heavy lines) and island chains (broken lines).

zontal movement. One of his sons, George Darwin, suggested in collaboration with Osmond Fisher that the Moon originated by being thrown off from the Pacific area of the spinning Earth or being drawn from it by the gravitational attraction of a passing star. This idea was extended by Fisher in 1882 when he suggested that the continents, which broke up at the time of the Moon's separation, would subsequently readjust their positions to the new shape of the Earth. This association of continental drift with the origin of the Moon dominated many of the later ideas of the twentieth century.

Early in the twentieth century two Americans, Frank B. Taylor and Howard B. Baker, independently and almost simultaneously, outlined their ideas on continental drift. Taylor, in 1908, suggested the phenomenon in order to explain the origin of our modern mountains (Figure 2), while Baker, in a series of articles between 1911 and 1928, used the match of these mountains on opposite sides of the Atlantic, although in his later articles he expanded his ideas to include other criteria (Figure 3).

Taylor, in particular, gave an elegant statement of the argument for large-scale continental movements, but most scientists regard the German astronomer, geophysicist and meteorologist, Alfred Wegener, as the real pioneer of the modern theory of continental drift. He was originally drawn to the idea to explain the ancient climates of the past (Chapter 5). Why should tropical ferns have grown in London, Paris, and even Greenland, yet glaciers have covered Brazil and the Congo at the same time? However, in his book *Die Entstehung der Kontinente und Ozeane*, published in 1915 (although the idea was presented in 1912), he drew evidence not just from the study of ancient climates, but from the whole range of the sciences. It was unfortunate that, in so doing,

3. Baker's 'Reconstruction Globe'

Between 1911 and 1928, Howard Baker joined the continents in this way so that the modern mountain chains would form continuous structures from one continent to the next.

he cited several examples which were incorrect or capable of quite logical explanation without recourse to a theory of continental movements. None the less, he invoked a tremendous body of evidence which formed the basis of extremely heated discussions, particularly during the 1920s (Figure 4).

Most of these debates centred on the 1924 English translation of Wegener's 1922 German edition, despite a further edition before his death while exploring Greenland in 1930. Many of the objectors merely expressed their complete and utter disbelief, with little attempt to justify their rejection of the concepts, but the majority of the opposition arose in one of two ways. Experts in one scientific field would either demonstrate errors in the details of evidence which Wegener had quoted and therefore conclude that the whole basis was incorrect, or they were even less logical and accepted the majority of Wegener's evidence from their own speciality but then disproved continental drift by quoting evidence from other sciences. Typical of the latter was the way in which many biogeographers accepted Wegener's statements that ancient plant and animal life clearly demonstrated ancient land connections between the continents (Chapter 4), and then said that as geologists did not believe in continental drift these former land connections (continents or huge land bridges) must have sunk beneath the waves of the oceans – although the geologists and geophysicists had proved almost 100 years before that this was impossible. This piecemeal approach by specialists gave the impression that the vast majority of the evidence presented by Wegener was wrong and it was very difficult to see that the undoubted errors in detail did not prevent the overall argument being highly significant, if only because it was drawn from so many independent scientific fields. However, the main reason for disbelief, or at least scepticism, arose because there was no known mechanism at that time which was capable of separating the continents in the manner that Wegener, or his supporters and predecessors, had suggested. After all, the continents and

---

4. Alfred Wegener's concept of the evolution of the continents (opposite)

This reconstruction was based on evidence from many scientific fields and is remarkably similar to our present understanding of the evolution of the southern continents (Figure 33, pages 92–3). The shaded areas indicate where shallow seas covered the continents.

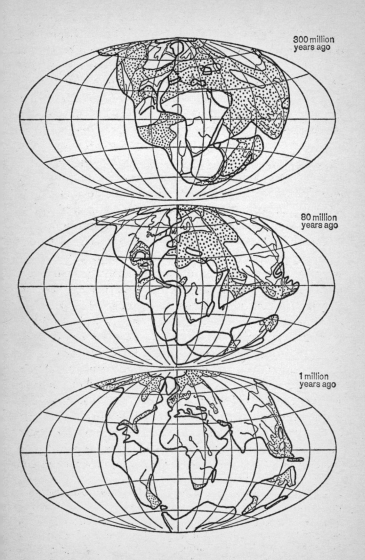

300 million
years ago

80 million
years ago

1 million
years ago

ocean floors are extremely solid rocks and tremendous forces must be envisaged to move even the smallest continent, Australia, which weighs some 500 million million million kilogrammes, through the oceanic floor. The most accepted mechanism at that time was still thought to be the effect of the separation of the Moon from the Pacific, but we now know that this is not the way in which the Moon originated, as it was formed at the same time as the Earth, 4,550 million years ago.

With hindsight it is easy to envisage these early opponents of continental drift as ultra-conservative diehards, but it is not really surprising that the majority of scientific opinion swung against continental drift during the two decades between the two World Wars and, indeed, there are still some observations which are not explained by large-scale continental movements alone. There were, however, some notable supporters. In particular the South African geologist, Alexander du Toit, undertook pioneer work in both Africa and Brazil. This work, extended later by Henno Martin, forms the basis of some of the best examples of the geological arguments for continental drift (Chapter 3). Another of the main proponents was a British geologist, Arthur Holmes, who pioneered the dating of rocks by radio-active methods and also propounded, in 1927, and more extensively in 1929, the mechanism of convection currents driven by radio-active heat within the Earth. He suggested that areas of the interior of the Earth became hot from the heat which Madame Curie found was released during the breakdown of the small amounts of radio-active elements contained in all rocks. The hotter areas would eventually rise, spread out near the Earth's surface, where the hot rocks would cool, and eventually sink back as cold, dense material into the interior of the Earth. Holmes suggested that these movements within the Earth could give rise to continental movements; the continents being carried across the face of the Earth like gigantic icebergs. Similar sorts of movements were almost simultaneously invoked by the famous Dutch geophysicist, F. A. Vening Meinesz, to explain his geophysical observations in the East and West Indies. Our modern ideas of these convection currents differ in some very important details from these original suggestions, but the basic driving force is essentially the same (see Chapter 9).

By the 1940s, therefore, there was a possible mechanism to account for continental drift and, as more and more geological evidence accumulated, more and more scientists became convinced of the reality of drift. This was particularly true of geologists in the southern hemisphere. The South African, Lester C. King and the Australian, S. Warren Carey, were particularly ardent proponents of drift and they obtained much new vital evidence. However, the main impetus to the wider acceptance of drift came in the 1950s with the rapid development of investigations in palaeomagnetism (the ancient magnetism of rocks) following the suggestion by the British physicist, Lord Blackett, of the application of new, extremely sensitive instruments capable of measuring this very weak residual magnetism. Using these intruments, the British geophysicist, S. Keith Runcorn, found that he could only explain his observations of the palaeomagnetism in rocks from Europe and North America in terms of continental movements (Chapter 6). With the extension of this work to other continents, mainly under the lead of British researchers such as Kenneth M. Creer in South America and Edward Irving in Australia, more and more people became convinced of the reality of large-scale continental movements. It was, however, the application of palaeomagnetic studies to the rocks of the ocean floor in the 1960s by Frederick J. Vine which finally led to a complete swing of scientific opinion towards the acceptance of the theory of continental drift (Chapter 7).

The fact of continental drift is not merely of academic interest: it is an extremely important economic factor to be considered in terms of mineral exploitation. The occurrence of valuable deposits on one continent can lead to the assumption that they might also occur on another, several thousand kilometres away. An example is the occurrence of diamond fields in West Africa and their counterparts in north-eastern South America. The oil-man is also concerned; although he is primarily interested in finding the geological structures which may contain oil, he has to bear in mind that economic quantities of oil, or natural gas, can only be produced if the rocks he is looking at were once in the right latitudes to allow the formation of oil in such quantities. The oil and natural gas fields of Europe and North Africa are only in existence because this part of the world was once lying much closer to the Equator.

Potentially more important is the development of the floors of the oceans which our knowledge of continental drift helps us to understand. With the mineral resources of the continents rapidly diminishing (on current estimates the world's present copper mines will be exhausted within the next 20 years) it will become increasingly necessary to tap the mineral resources of the ocean floors. This can only be done economically if we understand their geological history. However, before the start of economic exploitation, it is necessary, in this technological age, to have an inventory of the total resources of this planet, and yet two-thirds of the world's surface is hidden by the oceans whose potential can only be estimated if we know the story of their development.

On a much longer term, the understanding of the movements of the continents and ocean floors will lead to an understanding of the formation of mountains and the mechanisms of earthquakes and volcanoes. Such an understanding is essential to the prediction of natural disasters and possible methods of controlling such tremendous forces. Perhaps the most important aspect of all, however, is that the final acceptance of continental drift is resulting in an entirely new scientific outlook on the evolution of the surface of our planet which will lead us to a better understanding of the origin of the Earth.

If all the continents were once part of either a single huge land mass or two super continents, then it should be possible to see how the fragments would fit together. Unfortunately the jigsaw pieces have many straight edges, some minor parts are missing, and some have been added since, and the actual edges themselves are generally hidden below sea level or blurred by sediments from rivers. However, a glance at a map of the Atlantic (Figure 5) shows that, very roughly, the 'nose' of South America could fit into the curve of the West African coastline and that the coasts of both continents tend to parallel each other until the 'tail' of South America can be imagined to curl around the Cape of Good Hope. On the other hand, some scientists looked at this map and saw 'quite conclusively' that these two continents did not fit into each other 'by at least 20°'.

The difficulty is that it is impossible to test such a large-scale fit of parts of a three-dimensional sphere, the Earth, on the two-dimensional page of an atlas. Any map projection must involve some distortion in either shape or area. Unfortunately most globes, until recently, have also been unsatisfactory for this test as the majority were made by stretching quite large, flat triangular sections of printed maps on to a cardboard globe. Although modern globes are often made of plastic which has been formed over an accurately plotted mould, considerable caution must still be exercised to be sure that there is no distortion.

This technical problem of how to fit the pieces can be solved in various ways, but a more fundamental problem is to decide what are the boundaries of the continents; in other words, where are the edges of the pieces which we are trying to fit? What, in fact, is a continent?

Conventionally, the edges of the continents are taken to be the coast at sea level, but this is completely unsatisfactory for our purpose as it merely marks the surface level of water in the world's oceans at any particular time. During the last million years of the Earth's history, the amount of water trapped as ice at the Arctic and Antarctic has waxed and waned, varying the amount of water in the oceans and so shifting the position of the coastlines by several hundreds of kilometres and drastically changing their shapes. Obviously the true shape of the continents of the Earth lies far beneath the surface of these waters.

The first men to study the shape of the ocean floors were the mariners who put out a lead and line from their sailing ships – the length of line paid out being a measure of the depth of water. This was an extremely slow operation in deep water, but nevertheless, the main features of the ocean floors became known in this way before the end of the nineteenth century. It was found that the sea floor slopes very gently away from the land at first, forming the continental shelf, before plunging more steeply down the continental slope into typical oceanic depths of five to six kilometres (Figures 5 and 6). Modern techniques have considerably improved the detail of these early observations (some of these details will be discussed in Chapter 7) but they do not radically alter the picture.

It is clear that the continental slope is a very important surface feature of the Earth, and geophysical techniques, originally developed to detect the location of buried

Spitzbergen

Jan Mayen

Iceland

Azores

St Paul

Ascension

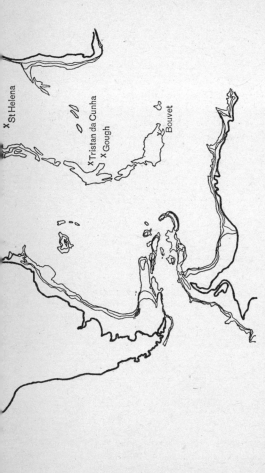

x St Helena

x Tristan da Cunha
x Gough

Bouvet

5. *The bathymetry of the Atlantic Ocean basin*

The closeness of the depth contours for 1,000, 2,000 and 3,000 metres illustrates the steepness of the continental slopes. The parallelism of the continental slopes is particularly striking on either side of the central and south Atlantic and their shape is reflected again at the mid-oceanic ridge by the presence of the 3,000-metre contour near the centre of the ocean.

minerals or of submarines beneath the waves, have shown that the rocks forming the continental shelf are the same as those of the continental land mass and quite distinct from those forming the deep floors of the oceans (Figure 6). One of these techniques, gravity surveying, shows that

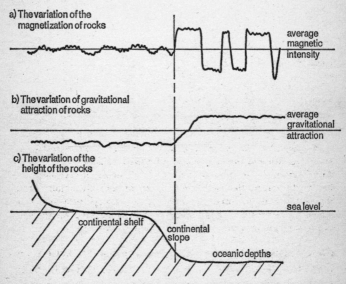

6. *The change of height and physical properties from continent to ocean*

The measurement of gravitational attraction shows that oceanic rocks are much denser than continental rocks and are also more strongly magnetized, accounting for the strong variations of intensity of the Earth's magnetic field over the oceans.

the continental rocks are much lighter than the true oceanic rocks, and that the boundary between the two types occurs at the continental slope. At their simplest (Figure 7), these gravimeters continuously weigh the same piece of material, the variations in its weight being a

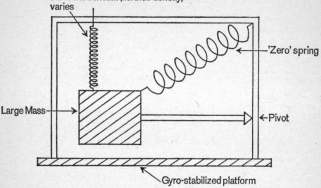

The extension of this Spring measures the change in weight of the suspended mass as the gravitational attraction of the rocks beneath (i.e. their density) varies

'Zero' spring

Large Mass

←Pivot

Gyro-stabilized platform

7. *A gravimeter for measuring the gravitational attraction of the rocks beneath and thereby determining their density*
The mass in the instrument remains constant so that variations in its weight record the variation in gravitational attraction. At sea the platform is stabilized by means of a gyroscope and any movements are measured by strain gauges, thereby allowing for the effect of the ship's movements on the instrumental records.

direct measurement of the gravitational attraction and therefore the density of the rocks beneath it. On land this material weighs as little as ten grammes, but at sea much larger weights are used and the instrument, developed in 1959, is mounted on a gyro-stabilized platform to try to reduce the effects of movements of the ship. Previously the only observation of gravity at sea had been made in the 1920s by the pioneer Dutch geophysicist, F. A. Vening Meinesz, who used a pendulum gravimeter in a stationary, submerged submarine, thereby avoiding the motion caused by the waves. Another shipboard instrument, the magnetometer, continuously measures the magnetization of the rocks beneath the ship. Once again, there is a distinct change from the weak magnetism of continental rocks to the much stronger magnetism of the rocks of the deep oceans as the research vessel passes across the continental slope (Figure 6). Quite clearly it is this slope which marks the boundary between the continents of the world and the true oceans.

Unfortunately, the geophysical observations are still few and it is not yet possible to plot the continental edges in sufficient detail to use these for fitting the continents. However, it is possible to use bathymetric charts for locating the continental slope and to attempt to fit the jigsaw edges on this demarcation. This is, of course, only a compromise and leads to some misfits where, for example, the Niger delta has been built out so rapidly that the shape of the continental slope has been distorted by its formation and the bathymetry does not, therefore, reflect the true edge of the continent. These misfits are few – indeed, one of the most remarkable things about the fit of the continents is that there are so few areas which must be excluded from the jigsaw.

As we saw in the previous chapter, several people have illustrated their ideas of how the continents fitted, but most of these were purely diagrammatic until 1958 when the Australian geologist, S. Warren Carey, tried fitting the continents bordering the Atlantic. To do this he plotted the continents on a large, 76-cm-diameter globe, and then made Perspex outlines of each continent which he could slide over the surface of his globe to test the various fits. Using this method, he demonstrated, very elegantly, the nearly perfect fit of Africa and South America (Figure 8(a)).

Surprisingly, Carey's reconstruction was regarded by many scientists as being too subjective and they required a more objective, mathematical demonstration. This was provided in 1965 by the English geophysicist, Sir Edward Bullard, who, with his colleagues, used a computer to report on the various fits of the continents bordering the Atlantic. The computer answer (Figure 8(b)) was in complete agreement with Carey's results and also showed that the best fit of all was obtained using the 1,000 fathom (2,000 metre) depth contour, corresponding to half way down the continental slope.

Since then, several people have used computers to test the fit of the remaining continents. This took longer as the locations of the continental slopes are not so well known for the other continents, particularly around the Antarctic, and the correlation between these slopes and the geophysical properties of the rocks on either side has only been studied in a few places. Nevertheless, it was found by 1969 that Antarctica, Australia, and India could all be fitted together extremely well, the fit of Australia and Antarctica being better than the fit observed across the Atlantic. There is not, however, a completely satisfactory fit of these three continents against south and eastern Africa because, at the moment, it is not known where the true continental edge of East Africa lies. It must lie farther from the coastline than the 2,000-metre contour, as sediments laid down on land have been located below this depth. The most likely, although not definite, geometrical fit (Figure 9(c)) can only be finalized after detailed matching of the pictures on the different pieces and, on all reconstructions, it is necessary to make corrections for areas which have moved relative to the main continental regions since their separation. This means, for example, that the South American 'tail' of Patagonia must be straightened, as must the Antarctic Peninsula which at present curves up towards it; in fact, much of West Antarctica must have moved relative to East Antarctica during the last 200 million years. Similarly, the fit across the North Atlantic is improved when Spain is rotated against France, closing the Bay of Biscay.

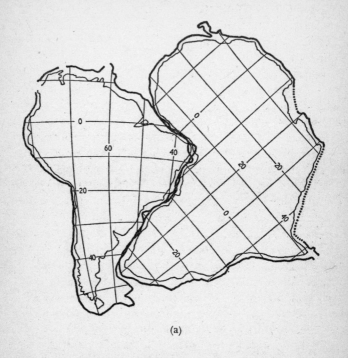

(a)

8. *Comparison of the graphical and computer fits of the continents bordering the Atlantic*

(a) S. Warren Carey's reconstruction used the continents plotted on spherical caps which could be moved over the surface of an accurately made globe into a position of best fit determined by eye which was then tested graphically (*above*).

(b) Sir Edward Bullard's reconstruction used a computer to move the continents into positions where there was the least overlap or gap between the pieces (*opposite*).

(b)

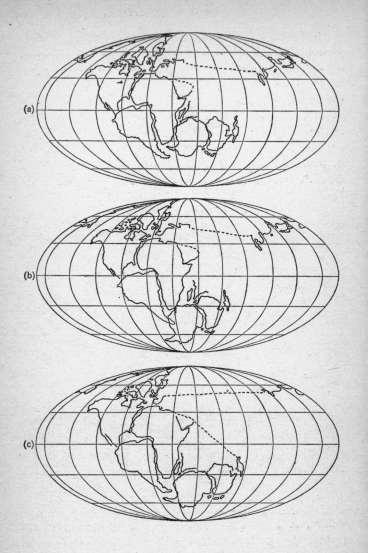

(a)

(b)

(c)

Although there are areas of the world where the continental edges are poorly defined, such as off East Africa, or have changed drastically during the last few million years, for example the Indonesian region, the areas where the continental edges are clearly defined fit remarkably well; so well, in fact, that it is impossible to explain the fit in any other way than that they once formed a huge continent which has subsequently been fragmented. However, we must now look at the picture on the jigsaw and see how this information helps to trace the past geography of our planet.

---

9. *The geometrical fit of the continents* (opposite)

Three main geometric fits have been proposed during recent years; (*a*) based on Smith and Hallam (1970), (*b*) Tarling (1970), and (*c*) this edition. There is agreement on the Atlantic fit, and the Australian–Antarctic fit, but there is still uncertainty about the fit of the other pieces, particularly in S.E. Asia.

When reconstructing the picture for various periods through geological time we have to match the continental pieces on the basis of either their prolonged similarity in geological development or the simultaneous occurrence of some specialized event. The evidence is, of course, restricted as we cannot peel off the younger layers of rock to reveal the full picture for any specific time, so we must rely on isolated fragments of old rocks which have either not been covered or whose cover has eroded away. Clearly we must be sure that the fragments we are using are the same age. This age correlation can be provided in two different ways – by using fossil plants and animals or by radio-active dating methods.

Fossils have been used for dating since 1811 when William Smith, the Father of Geology, found that any one layer of a particular rock contains a unique set of fossils. Although the same fossils can be found anywhere within the layer, identical groups of fossils do not occur in the younger overlying rocks or the older rocks beneath. This meant that once the time sequence of fossil assemblages could be worked out, rocks containing identical groupings could be dated by comparison with the sequence. We now know that this follows from Charles Darwin's concept of the evolution of species. Because new plant and animal species are being continually produced through the influence of environment on genetic changes it follows that at any one time the evolutionary assemblage of species must be unique. This method is often called relative dating as it dates a rock as older or younger relative to another, rather than measuring its actual age in millions of years (Figure 10). In practice, it is quite impossible, and unnecessary, to study detailed assemblages of fossils in order to date rocks. Generally some fossil species can be found which was evolving very rapidly at some particular time. The discovery of just one of these *index fossils* may be sufficient to date the rock it came from with very high precision. However, while this dating method is excellent within a single continent, there are few index fossils which have sufficiently wide geographical range to be suitable for attempting time correlation between different continents.

One other problem in using fossils for dating is that they can only be used for the last 570 million years of the Earth's 4,550-million-year existence. Although the first signs of life which can be found are over 3,000 million years old, these organisms did not have any hard parts until the sudden appearance of shells and skeletons 570 million years ago (Figure 10); the reasons for this sudden appearance are by no means clear. This means that we must use radio-active methods of dating for rocks from times before the existence of easily recognizable life forms, for time correlation, and also for establishing an absolute time scale in millions of years against which the fossil record can be compared.

Most rocks contain minute amounts of radio-active elements, such as uranium, potassium, and rubidium, which gradually decay to form other elements. As we know the rate at which this decay takes place, we can calculate the age of these rocks by measuring the amount of secondary, radio-genic element now present. The decay of radio-active potassium is illustrated in Figure 11. As the amounts of radio-genic elements are small, very

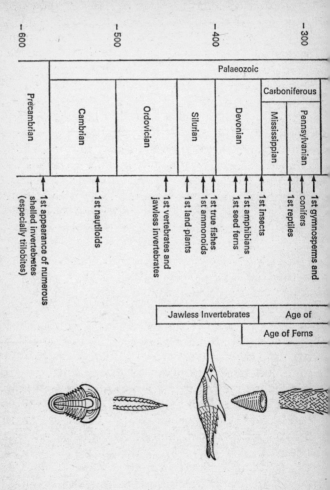

Time scale (millions of years):

| -600 | -500 | -400 | -300 |

Palaeozoic

Precambrian	Cambrian	Ordovician	Silurian	Devonian	Carboniferous	
					Mississippian	Pennsylvanian

- 1st appearance of numerous shelled invertebrates (especially trilobites)
- 1st nautiloids
- 1st vertebrates and jawless invertebrates
- 1st land plants
- 1st ammonoids
- 1st true fishes
- 1st seed ferns
- 1st amphibians
- 1st insects
- 1st reptiles
- 1st gymnosperms and conifers

Jawless Invertebrates | Age of

Age of Ferns

10. The stratigraphical column

From their fossil content geologists have been able to place rock[...] chronological order for the last 570 million years. Radio-ac[...] dating has determined the age of these sequences and dated rock[...] earlier times of the Earth's 4,550-million-year existence.

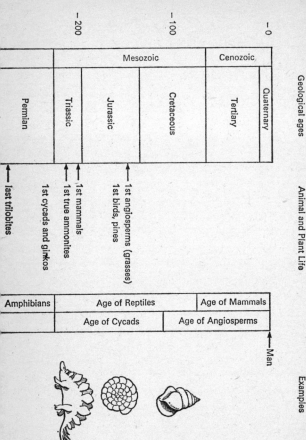

Age in millions of years

Geological ages

Animal and Plant Life

Examples

	Mesozoic			Cenozoic	
Permian	Triassic	Jurassic	Cretaceous	Tertiary	Quaternary

- 200
- 100
- 0

← last trilobites

↑ 1st cycads and ginkos

↑ 1st true ammonites

↑ 1st mammals

↑ 1st birds, pines

↑ 1st angiosperms (grasses)

Amphibians	Age of Reptiles		Age of Mammals
	Age of Cycads	Age of Angiosperms	

↑ Man

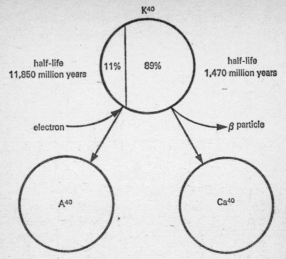

11. The decay of radio-active potassium

In a newly formed rock, just over one hundredth (0·0119 per cent) of the potassium will be radio-active and will break down to form either calcium or argon. This calcium is similar to ordinary calcium, but the argon produced differs from ordinary argon, so the amount of argon produced since the rock was formed can be measured. In 11,850 million years, half of the radio-active potassium which will eventually convert to argon will have done so, and in another 11,850 million years half of the remaining potassium will have changed, and so on—hence the rate of decay is measured as the half-life; the time it takes for half of the decay to take place. As we can calculate the amount of radio-active potassium originally present and know its decay rate, the measurement of the amount of radio-genic argon allows us to calculate the age at which the rock crystallized.

specialized equipment has been developed to measure them. At present the main difficulty with these techniques is to obtain rock samples which have retained all of their radio-genic elements and have not had them added to from gases or liquids which may have passed through the rock since its crystallization. Clearly we have here an extremely powerful technique for dating rocks, but we must be careful in comparing very old rocks as the error in measurement is still about 3 per cent so that, in an extreme case, two rocks both dated as 2,000 million years old, could differ by as much as 120 million years.

In view of all the problems in obtaining a good time correlation between rocks in different continents, it is not surprising that there have been many arguments on the interpretation of geological observations in terms of continental movements. However, with the advance in knowledge from both conventional geological methods and the improvement of radio-active dating facilities, we now have many examples of the matching of geological features. Of these examples, we shall look at one of detailed matching on radio-active evidence, and two of matching geological events common to continents which are now on opposites sides of the Atlantic. Then the remarkably uniform geological history of southern Africa and Brazil will provide an example of the general similarity of all the southern continents.

The first example relies on the application of radio-active dating methods to parts of South America and Africa. An area of rocks 2,000 million years old, in the western bulge of Africa, known as the Sahara shield, shows a structural 'graining', similar to the graining in wood but on a vast scale, which runs approximately north–south, but finally turns west and into the Atlantic. (This is particularly well shown in photographs taken from satellites passing over the Sahara.) Immediately to the east of these ancient rocks is another group of rocks which are only 550 million years old. The distinction between the rocks of these two different ages is extremely sharp and can be located running out to sea near Accra in Ghana (Figure 12). An expedition from the Massachusetts Institute of Technology went to Brazil to look for

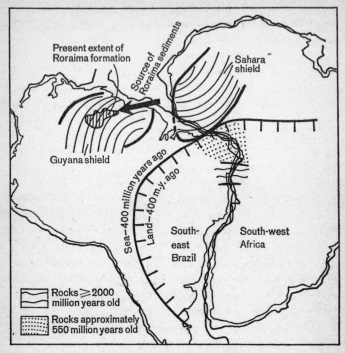

Present extent of
Roraima formation

Source of
Roraima sediments

Sahara
shield

Guyana shield

Sea – 400 million years ago

Land – 400 m.y. ago

South-
east
Brazil

South-west
Africa

☐ Rocks ≥ 2000
million years old

⦙ Rocks approximately
550 million years old

12. The geological match across the South Atlantic

The boundaries between rocks of different ages clearly match after
the continents are reconstructed, as do many geological features,
such as the shore-line of the Silurian sea, 400 million years ago, and
the Roraima formation (page 43). The rocks in south-eastern Brazil
and South West Africa have a remarkably similar history for the last
500 million years (described in the text and illustrated in Figure 14,
page 45).

a similar boundary in South America and were able to
find it near São Luis – in exactly the place expected after
fitting the jigsaw. Furthermore, detailed studies in the
older rocks show that the 'graining' in South American
rocks appears to match up with those of Africa. Detailed

radio-active dating, which is still continuing, has already shown more areas of identical age facing each other across the South Atlantic Ocean.

In the North Atlantic continents (Figure 13) there are three parts of a very large, ancient mountain system which can be seen running through Scandinavia, Scotland, and Northern Ireland; they are known as the

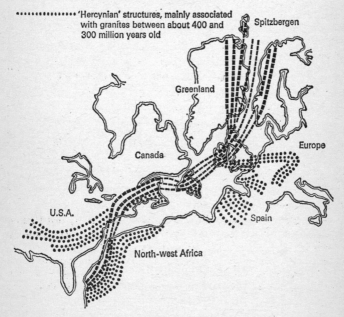

13. The geological match across the North Atlantic

If the continents are fitted together, the Appalachian–Caledonian Mountains form a continuous unit although the main phase of mountain building occurs at slightly different times in different places. It is not yet clear how well the mountain building in north-west Africa matches with the American end of this chain.

Caledonian Mountains from the earliest geological descriptions in Scotland. As this mountain system is over 400 million years old, most of the original mountains, once as high as the Himalayas, have been eroded away, exposing the very hard rocks (compressed while the mountains were forming) which now form the foundations of the Scottish Highlands. In Norway only the eastern edge of the system can be located, while in Greenland only the western edge can be traced. In other areas the situation is more complex. The Caledonian Mountains in Scotland are clearly a continuation of the Scandinavian-Greenland parts, but its further extension beyond Ireland appeared to be lost beneath the Atlantic Ocean. In Newfoundland an old mountain system comes out of the Atlantic which is identical, in most respects, to the Caledonian Mountains in Europe. This North American equivalent, the Older Appalachians, follows the eastern seaboard of the United States and eventually becomes lost beneath younger sedimentary rocks of the southern States. It is possible, however, that an extension of this same mountain system can be found in western Africa although our picture in that region, and to a lesser extent in America and Britain, is complicated by another mountain system which formed some 100 million years later.

Mountain chains, like the present Alps, Himalayas, and Rockies, are formed of rocks which were originally sediments. They were carried by rivers from a land mass and deposited in a trough which later underwent intense heating and pressure which converted them into extremely hard compacted rocks. (The processes which must operate to do this on such a gigantic scale are not really known but, in Chapter 10, we will discuss the latest ideas on these and other geological processes.) When rocks are eroded from the land, their debris is carried from the higher land by streams and rivers and during transport becomes more and more broken up, so that the sediments dropped by the rivers become finer as they travel farther away from their source. By studying the size and composition of particles in old sedimentary rocks it is possible to work out the direction and type of land from which they were derived. In Britain, we find that the source of many Caledonian Mountain sediments was a very extensive land mass which must have lain to the north and west where there is now the deep Atlantic Ocean. In North America, the sources of many Appalachian rocks lay to the south and east. To explain this before the acceptance of continental drift, geologists supposed that a continent 'Atlantis', must have occupied the present position of the Atlantic. This continent was thought to have sunk beneath the Atlantic waves. As we saw in the previous chapter, and will see in Chapter 7, there can be no question of the existence of a sunken continent in the Atlantic. By reconstructing the jigsaw we not only fit together the Caledonian Mountain chain, but also explain the sources of the sediments which formed it.

An even more striking example of the location of source material is the African origin of sediments of the South American Roraima Formation. In and around Guyana (Figure 12), this formation, more than half of which has been eroded away since its deposition over 1,000 million years ago, still covers over a million square kilometres and contains at least a million cubic kilometres of sediments which have been carried into South America from the north-east. The juxtaposition of Africa against South America offers an obvious source for these sediments and it is interesting to note that the diamonds, found at the bottom of the formation, get larger towards the Atlantic, while in West Africa similar rocks contain similar diamonds which continue to get larger towards their probable source in the Sudan.

Many other examples of this geological matching can be found between the continents bordering both the Atlantic and Indian Oceans. In fact, the geological development of all the southern continents is remarkably similar for the last 1,000 million years and can be illustrated by a brief glance at the last 550 million years of history revealed by the rocks in south-eastern Brazil and South West Africa (Figures 12 and 14).

For about 150 million years, from 550 to 400 million years ago, both areas were being severely eroded and worn down by rivers and winds until they were finally covered by sands and clays brought in by winds and rivers from neighbouring lands. Very shortly afterwards, both areas were glaciated (which we will discuss in Chapter 5) before developing forests which later gave rise to the coal deposits now mined in both regions. Following upon this, both areas were eroded for some 20 million years, bringing us up to 210 million years ago when both were covered by extensive beds of sand blown in from lands to the north. Subsequently both areas were covered by a shallow sea which left behind marine deposits which, in their turn, were overlaid by massive lava flows that covered over three-quarters of a million square kilometres in Brazil and even greater areas in southern Africa. These lavas, some only 100 million years old, were followed by a return of the sea and marine deposits of the same age are found in both areas. This final episode seems to mark the last event common to both areas and suggests that this was the shallow predecessor of the sea

14. The rock sequences in South-east Brazil and South West Africa (opposite)

The sequence of sediments and lavas on both sides of the Atlantic is almost identical and clearly suggests a common history although the areas are now separated by over 5,000 kilometres of deep ocean.

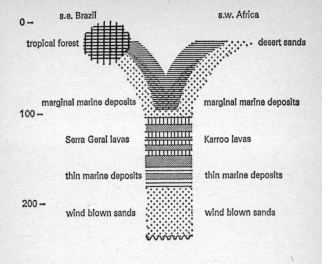

Age in millions of years

0 —

s.e. Brazil   s.w. Africa

tropical forest   desert sands

marginal marine deposits   marginal marine deposits

100 —

Serra Geral lavas   Karroo lavas

thin marine deposits   thin marine deposits

200 —

wind blown sands   wind blown sands

erosion — no deposits

300 —

coal   coal
ice erosion with   glaciation with
glacial deposits   glacial deposits

sands and clays   sands and clays

400 —

erosion — no deposits

500 —

old rocks   old rocks

which later deepened to form the Atlantic Ocean as the two continents separated (Chapter 8). Subsequently tropical forests have spread over the South American part and hot, arid conditions developed in South West Africa, marking the first stages of their independent development.

This is, of course, a simplification of a very detailed story in which even the minor events occurred simultaneously. Clearly the chance of exactly the same pattern occurring over such a long period of time in two quite separate areas is extremely unlikely.

4 ANCIENT LIFE AND ENVIRONMENTS

The fossil record of ancient plants and animals can tell us much about the conditions which existed in the past. All organisms develop a certain mode of life adapted to suit a particular environment. This may be as large as a continent or the sea itself, or just a few square centimetres of land. This means that, by mapping the distribution of different organisms in the past, we can locate and match the surroundings in which they existed. The interpretation of past environments based on their ancient flora and fauna must be done carefully as their distribution is controlled by the complex interaction of many factors about which we know very little – as demonstrated by Man's modifications of Nature which so often produce disastrous results.

The most distinguishable environments which can be mapped are marine and terrestrial, but in more detail we can, for example, map the changing depths of water which covered Britain some 450 million years ago from the nature of the sediments and the forms of life preserved in the rocks today (Figure 15). Some fossils can also be used to map even more precise environments, such as that of the coral reef.

Modern reef corals only thrive in clear, shallow sea water between 25° and 30°C (78–85°F), so that their distribution is a reflection of the occurrence of this narrow temperature range in shallow sea water. If we look at old coral reefs, the past location of the same precise environ-

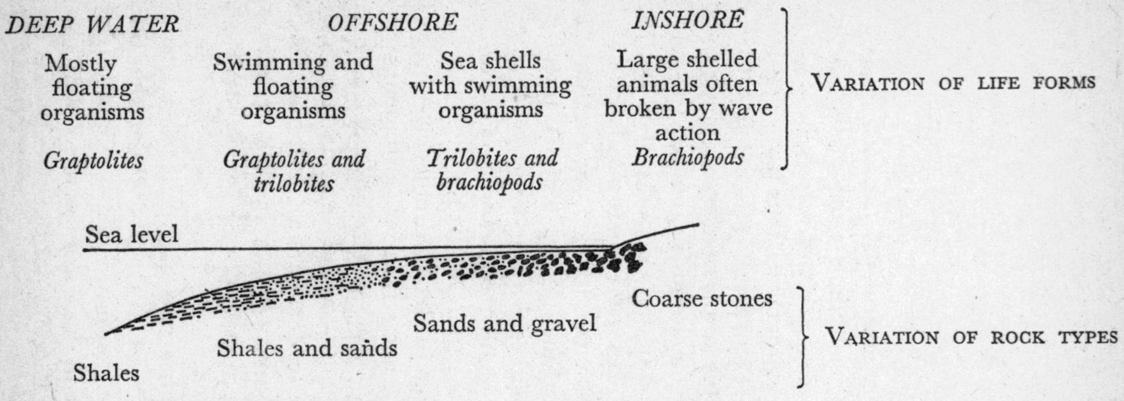

DEEP WATER	OFFSHORE		INSHORE	
Mostly floating organisms	Swimming and floating organisms	Sea shells with swimming organisms	Large shelled animals often broken by wave action	VARIATION OF LIFE FORMS
Graptolites	Graptolites and trilobites	Trilobites and brachiopods	Brachiopods	

Sea level

Coarse stones

Sands and gravel

Shales and sands

Shales

VARIATION OF ROCK TYPES

15. *The distribution of rocks and life in Britain 450 million years ago*

The relationship between the type of fossil life and the depth of the sea is very clear at this period. We know from recent observations of the ocean floor that great care must be taken before interpreting the occurrence of sands and clays in terms of the distance from the shore, so when reconstructing, *all* the factors available must be taken into account, e.g. the nature of the rocks and of the plants and animals which have been preserved in them.

ment can be traced. For coral reefs more than 100 million years old it becomes more difficult to be certain of the exact temperature range of the shallow sea water which they represent as the species were not the same as those existing today. However, we can still be sure that we are mapping restricted conditions which must be at least very similar to those existing on coral reefs today. These ancient coral reefs show that 350 million years ago, exactly the same precise conditions existed in both eastern North America and western Europe, and the limits of this environment can be matched if we reconstruct the continental jigsaw (Figures 16 and 18). It is even more striking that on both continents these reefs contain identical species of coral and other seashore and river life which clearly shows that these two shallow water areas must have been in direct communication, without any intervening oceanic or other hostile barrier.

By the end of the nineteenth century, naturalists had discovered sufficient examples of similar fossil species on different continents to conclude that there must have been very wide land connections for long periods of time between South America, Africa, India, Australia, and Antarctica. This interconnected land mass they called Gondwanaland (Figure 16) after the Indian kingdom where fossil plants, typical of all of these continents some 300 million years ago, were first described in detail. These plants, the *Glossopteris* type, were stunted in their growth and few species are known, suggesting that they were growing in a cold climate (see next chapter). Similarly, North America and Europe were thought to be connected to form a second land mass which they called Laurasia (uniting the St Lawrence area of North America with Asia, via Europe). At the same time as the impoverished Gondwanaland flora was growing, Laurasia had dense, luxuriant, tropical fern forests, which subse-

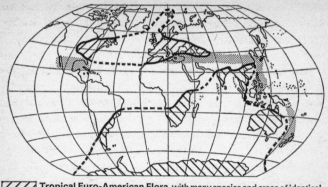

16. Different life environments 300 million years ago

The tropical northern and polar southern flora are shown with the continents in their present locations. The boundaries would clearly match on a continental reconstruction such as Figure 18, page 57. These two flora were separated by the Tethys in which marine fossils accumulated. This ocean was probably intermittently connected through the Mediterranean region to marine conditions in Central America at various times during the last 300 million years.

quently formed the coal beds of North America and Europe (Figure 16).

These early naturalists, and even some modern biologists, have explained these connections as very wide land 'bridges' which have since disappeared beneath the waves of the oceans. But it has been evident to geophysicists for over a century that such huge land bridges could not disappear without trace; all our knowledge of the oceans has still not revealed even the slightest trace of such sunken continental links, so that it *must* be concluded that these extensive land bridges never existed. Quite clearly, however, terrestrial animals could wander from, say, Africa to South America with no difficulty if these two continents were contiguous.

We must include whole populations of plants and animals in this kind of study because there is always the remote possibility that a few colonizers may have rafted, accidentally, across the oceans. In fact, we find many examples in the geological record of faunas and floras which are almost identical on different continents. For example, fossils of a whole range of large land reptiles some 200 million years old have been found in both East Africa and South America. There are many more such examples of animals which it is very difficult to imagine crossing even short expanses of salt water, yet freshwater fish, frogs, tortoises, and many others appear to have travelled very easily between South America and Africa until about 100 million years ago.

Perhaps the best example of the interplay of the movements of continents and of the animals upon them is afforded by the fascinating history of the marsupial and the placental mammals. In rocks about 100 million years old, the remains can be found in Europe, North America, and Asia of the first marsupial mammals. These marsupial ancestors, something like an opossum, rapidly spread to have a world-wide distribution by 70–80 million years ago, illustrating a high degree of freedom for animal movements between all of the continents. Shortly afterwards this freedom was lost as the continents separated and diverse marsupial species evolved on each continent. Meanwhile, the placental mammals evolved in Europe and Asia and spread rapidly into North America, suggesting that these regions were still connected. The new placental mammals were much more ferocious than the marsupials, being mostly carnivorous, while most of the marsupials were vegetarian. The placentals drove out, or quite literally ate out, the majority of marsupials in the northern hemisphere, leaving only those which had adapted themselves to environments which were hostile to the placentals. In South America, the marsupials lived quite happily in their isolation until some 30–40 million years ago when continental movements associated with continental drift produced the Central American link with North America which allowed placental mammals to invade from the north and to wipe out most of the South American marsupial fauna. Australia, however, remained quite isolated until the arrival of European Man, bringing with him many of his northern hemisphere animals who destroyed the native marsupial life until we are left today with only a handful of species of the flourishing Australian marsupial fauna of 200 years ago.

. Many marine animals, particularly shellfish, can only colonize efficiently along continuous shallow seas, usually closely following a coastline, so we can map the presence of these seas in the geological past from the occurrence of these fossil animals. For example, the species of fossil corals on opposite sides of Central America (the Bahamas and Baja California), show similarities and differences at different times during the last 50 million years indicating the opening and closing of direct sea connections between the Atlantic and Pacific as the Central American isthmus evolved. A great deal of similar evidence has been obtained from fossils in the sedimentary rocks that now form the Alps and Himalayas. This shows that, for most of the last 600 million years, there has been continuous water stretching from the Mediterranean through southern Asia, but north of India, into the Pacific (Figure 16). This ancient ocean, marked by both the sediments and the marine life preserved in them, has been recognized by geologists for a long time and named the Tethys (after the Mother of the Oceans, the wife of the Greek god Oceanus). We shall see in later chapters how this once extensive, deep ocean was converted into our present-day mountains as the southern and northern continents collided.

Our fossil evidence, therefore, shows a very distinct unity of the southern continents, South America, Africa, India, Australia, and Antarctica, and also of the northern continents – North America, Europe, and most of Asia. These two super-continents were separated by the Tethys Ocean, but the fossil record also shows that they were frequently in close contact with each other (Figure 18, page 57). This occurred, particularly, between northern South America, north-western Africa, south-western Europe, and south-eastern North America, often for very prolonged periods; for example between 400 and 250 million years ago when there appears to have been but one single huge horseshoe-shaped land mass, with fairly free migration between all land areas.

Chapter 8 will describe how some fossil marine organisms, ammonites, have allowed us to date, with high precision, the final separation of Africa and South America, but before discussing this evidence for the age of continental drift, we must look at the evidence, preserved in the rocks, of the changing climatic pattern on these super-continents.

Evidence of the past climates of the world is revealed in its rock formations just as clearly as the evidence for past forms of life. Graphic records can be found of former hot, salty deserts and glacial ice caps, and the significant factor of this evidence is that these signs of hot deserts are found in our present polar regions; in contrast, ice caps can be traced in present-day equatorial forest. Furthermore, these contrasting climates occurred simultaneously showing that the climatic zones on the continents at that time were in radically different positions from those of today.

There can be, clearly, two explanations: if we assume the continents have not moved, then the factors controlling the location of climatic belts must have changed; or, assuming continental drift, the climatic pattern could have been similar to today's but the continents must have been in different climatic regions. The present climatic zones of the Earth (Figure 17) are controlled by many factors – the distribution of land and sea, the presence or

---

17. The climatic zones of the world today (opposite)

These zones are clearly related to latitude because they are controlled by the angle of the Sun's rays to the Earth's surface. There are minor variations caused by the distribution of land and sea, e.g. the marked effect of the Gulf Stream on the climate of Western Europe, but these are small compared with the overall dependence on latitude.

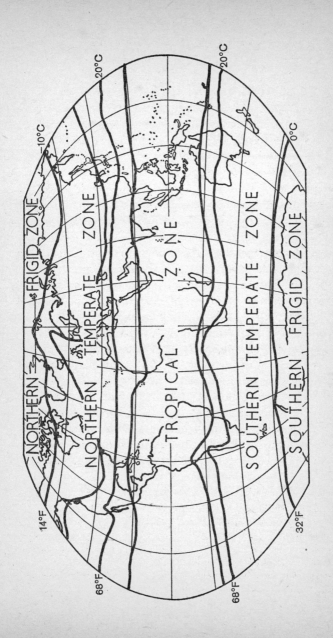

absence of mountains, etc., but above all else, the angle of tilt of the Earth's axis of rotation relative to the Sun. This means that we would expect some climatic changes during the geological past as mountains were formed and worn away, and seas advanced and retreated over the continents. But these changes would be small compared with those resulting from a change in latitudinal locations which could only occur if the continents moved or if the Earth's axis of rotation changed by a large amount relative to the position of the Sun. Such a change of the rotational axis would require the operation of such extremely powerful forces that it is possible that the entire planet would disintegrate under their action and they would certainly leave very strong traces in the geological record; yet there is no sign of such a traumatic event. It is possible, however, that the outer shell of the Earth may have slid as a whole over the interior of the planet. However, this explanation seems to require a more complex explanation than continental drift itself.

Between 100 and 200 million years ago, the Earth experienced a period of more even climatic conditions, generally warmer than today. This situation was almost certainly a reflection of the absence of mountains over most of the Earth and the very widespread, shallow seas on most continents at that time (Figure 33, pages 92 and 93). We must be careful, therefore, when studying a particular area to distinguish between the climatic changes caused by geological events, and the major changes which can only be explained by continental movements.

There is unequivocal evidence that between 250 and 350 million years ago there were hot deserts in the present north polar regions and glaciers in modern equatorial regions. This is particularly significant as the contrast between the two areas both then and now is very strong: it is easy to explain these observations using

areas of tropical coal forests 300 million years ago which some 20 million years later became vast hot deserts

areas of glaciation between 250 and 300 million years ago with arrows indicating known directions of ice movement

18. Polar ice and hot deserts between 250 and 300 million years old

The hot deserts illustrated here left behind vast sand dunes and salt deposits on Laurasia while simultaneously a polar ice cap was present in Gondwanaland. The glacial deposits and ice directions shown above are not all contemporaneous, being some 300 million years old in Brazil and about 250 million years old in Australia and Antarctica. During this 50-million-year period, Laurasia was drifting within the equatorial zone so that the desert belt gradually changed position, although the deserts and latitudes shown here are for about 250 million years ago.

continental drift, but extremely difficult to do so by any other method.

During this period tremendous glaciers swept over south-eastern South America, central and southern Africa, India, Australia, and Antarctica (Figure 18). This glaciation lasted for over 50 million years in some areas and was almost entirely in the form of huge ice sheets, similar to those in Greenland and Antarctica today. They scraped and plucked frost-shattered rocks off all of the Gondwanan continents and later deposited them as the ice eventually melted. Several hundreds of metres of this glacial debris can be found today on all of the Gondwanaland continents and the depth may be as much as 1,000 metres, as in Brazil, where it covers over 4 million square kilometres. These ice sheets were much more persistent and each covered larger areas than the Pleistocene sheets of the last Ice Age when, at times during the last 2 million years, north polar ice reached as far south as London, New York, and St Louis, leaving behind only some 100 metres of glacial debris over much of northern Europe and America. The extent of the Gondwanan glaciation indicates that it must have been a polar ice sheet, particularly as most areas affected were low lying, for mountain glaciers, such as those in the Alps and Rockies today, could not possibly account for the great volume of glacial debris, although there was some minor mountain glaciation at this time, for example in parts of eastern Australia.

The glaciation of South America and Africa commenced just before that of Madagascar and India which were iced over just before Antarctica and Australia. None the less, on the present distribution of the continents containing evidence of these ancient glaciers, it would appear that almost continuous ice must have stretched nearly simultaneously from the present South Pole to north of the

Equator, implying that, in fact, the whole world must have been covered by ice at this time. As we shall see, this was certainly not the case, but if we reassemble the southern continents into Gondwanaland (Figure 18), and allow the super-continent to traverse slowly through the southern frigid zone of the Earth, then the southern ice cap would appear to progress from one side of this continent to the other. Not only does this interpretation explain the different ages for the onset of glaciation across Gondwanaland, but it positions the climates marginal to the ice cap in their proper latitude for that time, so that the final stages of glaciation in Australia were contemporaneous with the appearance of hot deserts in eastern South America and northern Africa, then some 20° from the Equator. On their present distribution, these continents, being almost antipodal, would, of course, be expected to have been experiencing similar climates to Australia.

There is, however, even more evidence for the validity of the Gondwanan reconstruction from the debris and other signs of glaciation which still remain. It is possible to trace the direction of movement of ancient glaciers from the scratch marks left behind as they dragged boulders over the land surface (Figure 18). Furthermore, a study of these boulders can indicate their source as an extra check of the direction from which the glaciers flowed.

All the glaciers in Brazil at this time flowed from the east, where now there is the Atlantic Ocean, and in South West Africa several sets of tracks can be followed crossing what was then higher ground and passing westwards out into the Atlantic. The Brazilian glaciers carried debris from a continental area and this includes rocks known in South West Africa, but unknown in neighbouring areas of South America. Similar testimonies of the continental source of glacial debris carried from areas where there are now only oceans are found throughout Gondwanaland. In the extreme south of Africa, continental rocks came from the south; in south-eastern Africa, near Durban, several hundred metres of glacial rocks came from where the deep Indian Ocean is now only a few kilometres away; and in Australia, near Adelaide, glaciers brought great quantities of continental rocks from the direction of the present-day Australian Bight. It is only by reconstructing these continents into a single Gondwanaland that we can provide the source of these gigantic glacial deposits.

At this time in the present continents of the northern hemisphere the luxuriant tropical forests mentioned in the previous chapter were giving way to extensive hot, sandy deserts in which shallow seas quickly evaporated forming thick deposits of salt which can be found over much of North America and Europe (Figure 18) stretching as far north as the Canadian Arctic Islands and northern Greenland. These evaporites now form valuable sources of salt and potash and the geological structures associated with them often form the traps in which oil and natural gas has accumulated, such as those of Texas and the North Sea.

This contradictory evidence from the northern and southern continents clearly shows that all the continents must have been in radically different positions to each other 200–300 million years ago but the climatic belts are very broad and modified by local distributions of land and sea so that the exact location of the continents cannot be defined with this evidence alone. We can gain more precise information of their relationships from the study of the magnetism of their rocks.

So far our evidence has been interpreted in terms of large-scale continental movements although, in some cases, other interpretations have been possible. More recent research has developed methods which can actually measure the positions of the continents in the past. For this reason, the story told by the magnetism of continental rocks, discussed here, and those of the ocean floors, discussed in the next chapter, have been the main cause of the swing towards the scientific acceptance of continental drift.

Rocks containing iron minerals become magnetic in one of two ways. Igneous rocks, such as granites and volcanic lavas, are formed from once molten rocks; lavas erupting from volcanoes are just over 1,000°C (1,800°F). These very hot rocks are non-magnetic as heat destroys magnetism, but after they have solidified and continued to cool below 600°C (1,100°F), the iron mineral particles within them become magnetized in the direction of the prevailing Earth's magnetic field (Figure 19). This high-temperature magnetization is very stable and is 'frozen' within the lava, so that its direction is not affected by subsequent changes of the Earth's magnetic field. This means that igneous rocks can retain a record of the direction of the geomagnetic field at the time they were formed. This phenomenon was first observed at Mt Etna in the mid-nineteenth century, and many subsequent observations of historical eruptions and laboratory experiments have

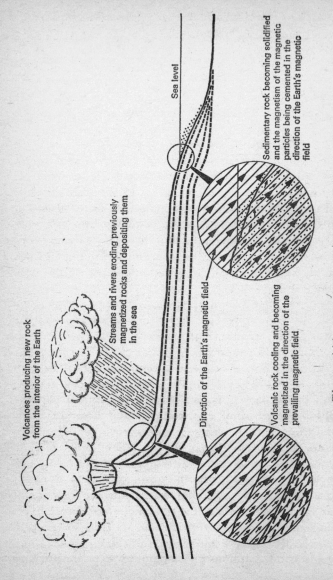

Volcanoes producing new rock from the interior of the Earth

Streams and rivers eroding previously magnetized rocks and depositing them in the sea

Sea level

Sedimentary rock becoming solidified and the magnetism of the magnetic particles being cemented in the direction of the Earth's magnetic field

Direction of the Earth's magnetic field

Volcanic rock cooling and becoming magnetized in the direction of the prevailing magnetic field

19. The acquisition of remanent magnetization by rocks

confirmed the fact that in this way the direction of the ancient geomagnetic pole can be preserved in igneous rocks.

Sedimentary rocks, such as sandstones and shales, are more complicated (Figure 19). They are mostly formed of particles which have been eroded from pre-existing rocks by the turbulent action of streams and rivers. The eroded material is carried away and gradually deposited along river banks, in lakes and eventually on the sea floor. The iron mineral particles in it are already magnetic as they have been eroded from previously magnetized rocks, but while they are being swept along by the currents, they lose their alignment until they are deposited as a wet slurry and are then free to realign themselves in the new direction of the Earth's magnetic field. As additional sediments are deposited on top of them they gradually dry out and become cemented into a hard sandstone or shale. This complex process of turning wet sediments into compact rocks, diagenesis, literally cements the magnetic particles parallel to the magnetic field and any iron minerals which form within the cement will also become magnetically aligned.

This means that both kinds of rock, igneous and sedimentary, can contain a magnetization which reflects the direction of the geomagnetic pole at the time that the rock formed. Unfortunately not all rocks are able to keep their original magnetic direction throughout geological time. In some it decays and is lost, but fortunately some rocks, particularly dark, igneous rocks and red sandstones, retain enough magnetization for it to be traced by various laboratory methods. From detailed studies of this palaeomagnetism of rocks the history of changes in direction and intensity of the Earth's magnetic field is gradually being built up.

At present the north magnetic pole is in northern

Canada, well away from the geographical (rotational) pole and one of the earliest results to emerge from the study of palaeomagnetism was that, over the last few

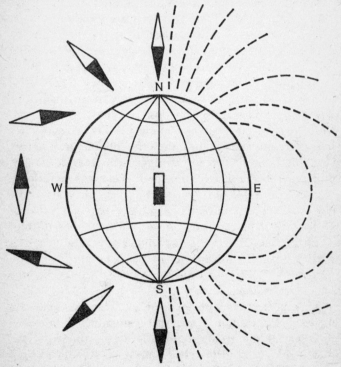

20. *The relationship between magnetic inclination and latitude*

The average Earth's magnetic field can be imagined as a simple bar magnet at the centre of the Earth aligned along its axis of rotation, so that the inclination of a suspended magnetic needle varies from vertical over the pole, to horizontal at the Equator. (The relationship is given simply by the formula that the tangent of the inclination is twice the tangent of the latitude.) Measurements of the inclination of the magnetization in rocks can therefore be used to determine the latitude in which the rocks acquired their magnetization.

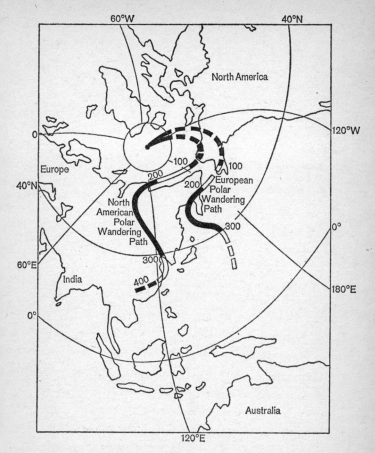

21. Polar wandering curves for Europe and North America

A curve can be produced for each continent by plotting the measured position of the pole (continuous line) and estimated position
(broken line) through geological time; marked in millions of years.
These curves are still incomplete but their separation is the same
width as the present Atlantic, until about 100 million years ago; they
would be identical for these earlier times if North America was
joined to Europe.

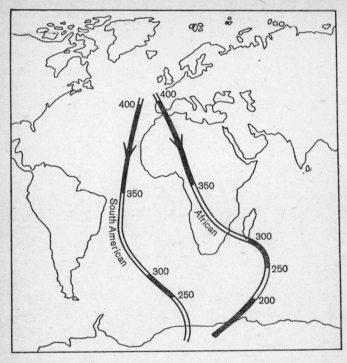

22. *The fit of Africa and South America to reconcile their polar wandering curves* (above and opposite)

(a) The polar wandering curves for the two continents in their present position (time in millions of years).

thousand years, its location changes and its average position coincides with that of the rotational pole. This relationship must be true for earlier times as the ancient latitudes which can be determined paleoclimatically are the same as those determined palaeomagnetically. The latitude in which the rock was magnetized can easily be calculated (Figure 20) from the angle of dip from the horizontal of the magnetic direction within the rock. The

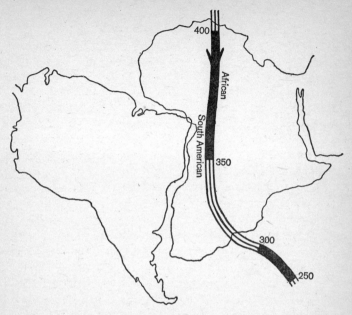

22. (b) Moving the continents to match their polar wandering curves produces, for a period of at least 150 million years, a fit identical to that required by the geometrical observations.

study of older and yet older rocks from any one continent shows that the average magnetic pole, and therefore the rotational pole, for different ages occurs in different positions and by joining these calculated pole positions for successive ages, we can plot a 'polar wandering curve' for that continent.

The polar wandering curves for North America and Europe (Figure 21) were the first to be studied and several curious features became apparent. Both curves were similar in shape, but before 70–100 million years ago the North American curve lay to the west of the European. This presented a situation where we had either to recognize two separate magnetic poles, which would also mean accepting two rotational poles, or to adjust the position of the different continents until the magnetic particles in their rocks pointed to the same pole at the same time. Subsequent work has confirmed this picture and shown that for a period of at least 200 million years (between 300 and 100 million years ago) the two curves are separated by exactly the width of the present Atlantic Ocean and, therefore, the only way to reconcile the pole positions is to move the continents together so that there is no vestige of the North Atlantic during this period.

We can, therefore, adjust the continents by moving them so that their polar wandering curves correspond for the geological period being investigated (Figure 22). Reliable data is still rather sparse for many of the continents, but from work still in progress it can already be seen that all of the southern continents have broadly similar polar wandering curves from at least 600 million years ago until between 200 and 100 million years ago.

The adjustment of the continents which best fits the present palaeomagnetic evidence is exactly that required to explain their ancient climates. For example, African data shows the magnetic pole located in the Sahara 500 million years ago, and the palaeoclimatic evidence, discovered while exploring for oil and gas in southern Algeria, shows that this region was then being glaciated; 200 million years later, the polar ice cap had moved on to South Africa, as we saw in the previous chapter, and the magnetic pole was also there at that time. In contrast, the northern continents were well away from the poles 300 million years ago – exactly as we had deduced from the age of their hot desert sandstones and salt deposits. The palaeoclimatic and palaeomagnetic evidence is, therefore, in excellent agreement and neither observation can be explained without continental movements occurring in precisely the same way, and the position of the continents, for example 100 million years ago (Figure 23), is in excellent accordance with the evolution of the continental patterns, as discussed in Chapter 8.

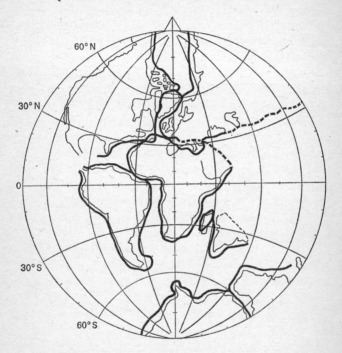

23. The continental configuration 60–100 million years ago using palaeomagnetic evidence alone

There is not yet sufficient data to reconstruct the continents at any one precise time, and it is possible that small continental movements may have gone on during this 40-million-year period, but the data shows that Africa–India and Australia–Antarctica were still united, while South America was just rotating away from Africa. In the north, the Labrador Sea was already open, but the North Atlantic, between Greenland and Norway, had not yet begun to form.

There is, however, another very important aspect of this study of the magnetism of rocks, the full significance of which has only recently been recognized. In the late nineteenth century it was observed that some rocks are magnetized in exactly the opposite direction to that expected. The magnetization of these rocks points towards the south magnetic pole. The discovery of these 'reversely' magnetized rocks in different parts of the world led to the suggestion in the early twentieth century that the Earth's magnetic field may change polarity, the north magnetic pole becoming the south and vice versa. This was confirmed by further studies in the 1950s which showed that all rocks of the same age are of the same polarity. It has been shown that these reversals occur throughout the history of the Earth, but during the last 70 million years they appear to have been more frequent, taking place at least once or twice every million years.

As the Earth's magnetic field is caused by electrical currents generated within the deep interior of the Earth (Figure 34, page 102) these reversals were studied in more detail to try to understand their implications for the internal structure of the Earth and also for their possible use to obtain accurate geological time correlations. This

---

24. *The periodicity of reversals of the Earth's magnetic field during the last 4 million years* (opposite)

Exactly the same polarity of the Earth's magnetic field can be found in volcanic rocks dated by radio-active methods and in sedimentary rocks dated by their fossil content. As the methods of magnetization and of dating are both different, the precise agreement shows that the Earth's magnetic field does change polarity. The sedimentary rocks have been dated using foraminifera which identify different time zones, identified by Greek letters. The times when the Earth's magnetic field had the same polarity as today are shaded and the different polarity changes are classified into epochs (Brunhes, etc.) within which briefer events occur (Jaramillo, etc.).

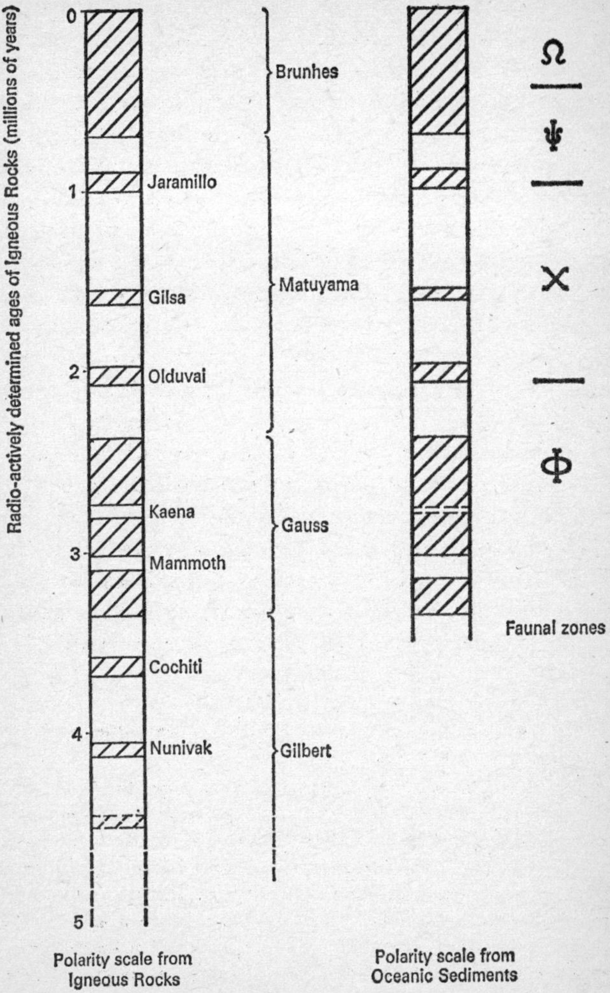

Radio-actively determined ages of Igneous Rocks (millions of years)

0

1 — Jaramillo

Gilsa

2 — Olduvai

Kaena

3 — Mammoth

Cochiti

4 — Nunivak

5

Brunhes

Matuyama

Gauss

Gilbert

Ω

ψ

×

Φ

Faunal zones

Polarity scale from
Igneous Rocks

Polarity scale from
Oceanic Sediments

research has shown that a reversal takes place during a period of a few thousand years and that the 'normal' or 'reversed' polarity is then maintained for periods varying from 100,000 years to 50 million years. The more recent reversals of the Earth's field have been studied in most detail as the palaeomagnetic observations for these are easiest to make and the young igneous rocks can be very accurately dated by radio-active methods. Using these igneous rocks, a table of polarity changes of the Earth's magnetic field has been built up for the last 4 million years (Figure 24). This table has now been confirmed by fossil-dated sediments cored from the deep oceans. These cores (described in the next chapter) are usually some 10–15 metres long and so give a continuous sedimentary record of magnetic changes during the last 3 million years (Figure 24).

One of the fascinating side lights of the study of these sedimentary cores is that many fossil species were found to disappear about the time of a reversal and new species appeared shortly afterwards. This marked increase in the rate of evolution is thought to be related to an increase in the amount of harmful radiation from the Sun which would be able to reach the surface of the Earth during these times of polarity change when the Earth's field would be very weak and consequently not be giving its usual protection. (The Earth's magnetic field has been weakening for at least 150 years and at the present rate of decrease would disappear in about 2,000 years. So perhaps we are now living through the start of a period of reversal after which the Earth's field will rebuild itself in the opposite direction.)

Polarity changes in earlier times are not so well documented. We know that reversals were still very frequent during the period from 5 to 100 million years ago, but these are more difficult to date as the length of each polarity period is about the same as the error in radio-active dating of rocks of this age. Therefore, for this period, it is difficult, with present techniques, to define the exact time of each polarity change. However, a table of less frequent, older polarity changes is being established. The next chapter will describe how these tables of reversals of the Earth's magnetic field provide the key to discovering the history of all our ocean basins.

Two-thirds of the Earth's surface is covered by oceans and these must have been transformed as the continents were moving into their present positions; therefore, we would expect to find evidence of these movements in the rocks of the ocean floors some 4 to 5 kilometres beneath their surface.

Until the late 1950s one of the major features of the Earth's surface, a mountain ridge 3 kilometres high and hundreds of kilometres wide, lay unknown beneath the waves. Although the central Atlantic was known to be shallower than the ocean on either side it was not realized that the line of mid-oceanic islands from Iceland to Tristan da Cunha (Figure 5, pages 24 and 25) forms only a small part of a 80,000 kilometre long ridge system. In the Atlantic the centre of this ridge is marked by a rift valley, some 2,000 metres deep and about 50 kilometres wide, which overlies the location of the majority of all earthquakes within the Atlantic Basin. By 1953, instruments for measuring the occurrence and location of earthquakes, seismometers, had been developed sufficiently to detect the presence of a line of these oceanic

---

25. *The location of the oceanic ridges, fracture zones cutting them, and the deep oceanic trenches* (opposite)

The centre of the ridge system is shown by a mid-oceanic line which is displaced by a large number of fracture zones, shown as fine lines. The oceanic trenches (discussed in Chapters 9 and 10) are shown by thick lines, most of which border the Pacific.

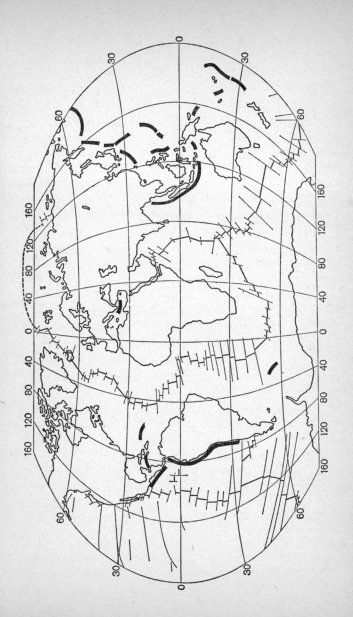

earthquakes extending from the central rift valley in the Atlantic all the way round Africa, through the central Indian Ocean and finally into the Gulf of Aden. The position of these earthquakes clearly suggested that the central rift valley, and, therefore, the ridge itself, formed a continuous system through both of these oceans.

Between 1956 and 1960, American and British ocean-ographic expeditions used depth recorders to trace this ridge system (Figure 25), not only exactly along the centre of the Atlantic and Indian Oceans, but also mid-way between Australia and Antarctica finally to link with a ridge running northwards through the eastern Pacific. They found that the centre of the ridge is not always marked by a rift valley and that whole segments of the ridge system are offset from each other by tens and sometimes hundreds of kilometres along tremendous fractures in the Earth's surface (Figure 25). These frac-ture zones can be traced for very long distances, some-times exceeding a thousand kilometres. Between the off-set centres of the ridge, these fractures are the sites of many of the oceanic earthquakes. These and other fea-tures of this fundamental ridge system will be discussed in Chapter 9. The discoveries made by 1960 were already significant for the continual drift argument as the frac-ture zones demonstrated quite unambiguously that very large scale movements of parts of the Earth's surface had taken place, and in the Indian and Atlantic Oceans the parallelism of the ridge with the corresponding conti-nental slopes further substantiated the geometrical evid-ence.

Rock samples dredged from the ridge show that its craggy relief (Figure 26) is formed by igneous rocks, as had been suspected from the volcanic nature of the mid-oceanic islands. Photographs of the sea-floor confirm that the centre of the ridge is formed mainly of lavas, with

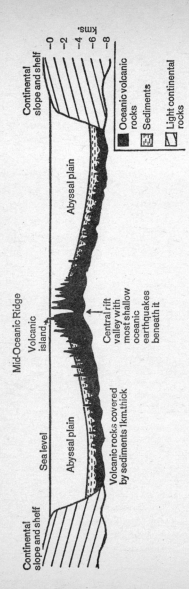

26. *Diagrammatic cross-section of a mid-ocean ridge*

The rugged nature of the volcanic rocks can be seen disappearing beneath a blanket of sediments on either side of the ridge axis.

little or no sediment for about 100 kilometres on either side, beyond which the volcanic peaks gradually disappear beneath the sediments. These thicken to form a layer, generally 1 kilometre thick, overlying the igneous floor as far as the edge of the ocean basins, the continental slope.

The top of these sediments can be dated from the fossil content of samples raised with a core barrel. This is a weighted steel tube released just above the ocean floor so that it drops vertically, usually for some 10–15 metres but sometimes up to 30 metres, into the sediments. As this depth of deposit only represents the most recent 2–3 million years this method is of little use for dating the ocean basins except in a very few localities where much

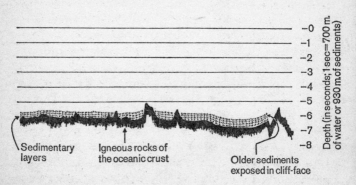

27. *A sketch of an echogram, illustrating the thickness and layering of sediments on the igneous ocean floor*

Pulses of sound reflecting from layers within the sediment produce this type of picture of the sedimentary layers as the ship passes over them. Except for a band along the centre of the oceanic ridges, these sediments obscure the igneous rocks, although occasional volcanoes still project through them. The top layers of the sediments are almost always those laid down during the last few thousand years, but in about six localities the older sediments are exposed, as in the cliff face illustrated on the right, from which samples can be obtained using core barrels.

THE OCEAN FLOORS 79

older sediments are known to be at the surface (as described below). Our knowledge of the nature and distribution of these older sediments has been radically improved during the last few years by the use of the 'sonic profiler'. This instrument works in the same way as a depth recorder which determines the distance of the sea floor beneath a ship by timing an echo bounced off it, but the sonic profiler uses a more powerful, lower frequency sound source, so that the pulse of sound waves actually penetrates the sediments and is reflected back to the surface detectors by the different layers. The echograms (Figure 27) produced in this way can give a continuous picture of the layers of sediments beneath the ship.

In several parts of the oceans, some very distinct reflections can be traced over several thousand square kilometres, particularly in parts of the Atlantic and Pacific oceans. One of these reflectors, termed horizon A, appears to mark a change in the nature of the sediments, with very smooth bedding below and more turbulent deposition above. However, the significance of these layers was difficult to evaluate until the age was known.

Detailed examination of numerous echograms revealed that, in a few places, the lower sediments are exposed at the surface of the ocean floor by either a recent fracture in the underlying igneous rocks (Figure 27) or by exceptional submarine erosion which has removed the younger sediment. Very careful sampling was made in these areas to collect sediment for dating. This was extremely difficult as it involved lowering a core barrel from a drifting ship through about 5 kilometres of water with its varying currents and dropping it into a small area of exposed lower sediments on the side of a cliff. None the less, samples were eventually obtained which showed that one of the reflectors, horizon A, was formed of limestones some 70 million years old and another horizon, β, which is

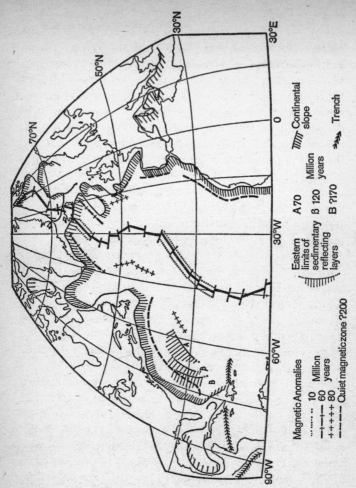

28. *Some of the evidence for the young age of the North Atlantic basin*

The age of the oldest sediments at any point on the ocean floor always increases away from the centre of the ocean as shown here for the sedimentary reflecting layers and confirmed by drill cores taken right across the basin. A similar increase in age is also shown by the magnetization of the igneous rocks underlying the sediments (described later in this chapter).

close to the bottom of the sedimentary sequence, was 120 million years old. The full significance of the distribution of these and other reflecting layers then became apparent for they are only found well away from the ridges and the area of each horizon decreases with age (Figure 28). This dating showed that the age of the oldest sediments increases with distance from the ridge and that the oldest of these rocks are very young compared with the rocks of the

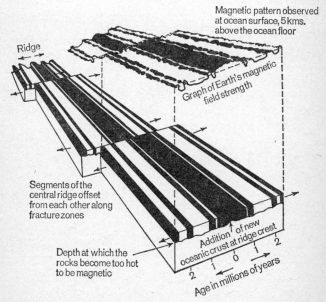

29. *The magnetic patterns of the ocean floor and at the ocean surface*

The observation of a strong magnetic field at the ocean surface corresponds to the igneous rocks of the ocean floor beneath being magnetized in the same direction as the present Earth's field (both shown as black); while lower values of the field correspond to reversed magnetization in the rocks beneath (shown as white). The pattern becomes confused at fracture zones, but can be matched away from the disturbed regions.

continents which are sometimes over 3,000 million years old.

This age distribution was finally proved during 1968–70 when a series of holes were drilled across the Atlantic and Pacific during a collaborative effort by four American oceanographic institutions (although this work has since been greatly expanded as the Deep Sea Drilling Project). This project included drilling right through the sediments and sampling basal sediments which were found to systematically decrease in age towards the centre of the ocean floor.

Sampling of the igneous rocks of the floor beneath the sediments is difficult or impossible except on the central parts of the ridge and most rocks obtained from these areas by dredging could not be dated accurately by radio-active methods as this surface material had already been subjected to chemical reactions with the sea water. However, ages obtained from the least corroded samples of the ridge crest were usually less than 10 million years old. Quite unexpectedly it is the strong magnetism of these igneous rocks which has recently allowed us to date them, not only where they are exposed on the ridge crest, but throughout the ocean basins.

---

30. *The origin of the magnetic pattern of the ocean floor* (opposite)

(a) Hot rocks rise from the interior to the Earth's surface at the centre of the ridge.

(b) As they cool, they are magnetized in the direction of the prevailing Earth's magnetic field.

(c) New molten rocks rise as the earlier rocks are displaced to each side. The new rocks cool and are magnetized in the Earth's magnetic field which has now reversed its polarity.

(d) Further rocks are added and are magnetized in their turn. The displaced rocks preserve a record of the previous polarity changes which can be recorded by ships on the ocean surface.

a)

Oceanic Ridge crest

Oceanic crust

Molten rocks rising to the surface at the ridge crest

b)

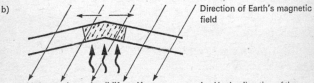

Direction of Earth's magnetic field

Molten rocks cool, solidify and become magnetised in the direction of the prevailing field

c)

New rocks rise as older rocks are displaced to each side. These in their turn cool and become magnetized, but the field has now changed direction and their magnetization is therefore opposite to that of the earlier rocks

d)

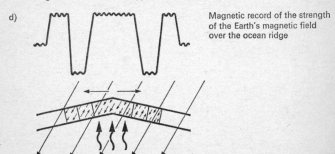

Magnetic record of the strength of the Earth's magnetic field over the ocean ridge

As more rocks rise and become magnetized, the older, displaced rocks preserve a record of the previous polarity changes of the Earth's magnetic field and their magnetization can be recorded by ships on the ocean surface.

Variations in the strength of the Earth's magnetic field over the oceans have been mapped for many years from measurements made with magnetometers developed for the aerial detection of the magnetic disturbance caused by submarines. These variations are much larger over the oceans, indicating that the oceanic rocks are much more magnetic than continental rocks (Chapter 2). It was not known why these variations should be so extreme, although they proved useful in 1959 when it was found that the centre of the ridge, whether marked by a central rift valley or not, is always associated with a particularly strong magnetic anomaly. This means that where several valleys are present, as in parts of the Indian Ocean, the central valley could be recognized and where no valleys exist, such as in Eastern Pacific, the displacement of segments of the ridge along fracture zones could still be measured from the offset of the central anomaly.

At the time that this association was discovered, extremely detailed surveys of the magnetism of the Pacific floor were being made at the Scripps Institute of Oceanography of California. They found that the magnetic variations are not random but form simple patterns of long narrow strips where the Earth's magnetic field is alternately much higher and then much lower than average. These strips (Figure 29) vary in width, up to 30 kilometres, and extend for hundreds of kilometres between

31. The rate of opening of the oceans from their magnetic patterns (opposite)

As a ship travels away from a ridge, it crosses the same polarity zone at different distances in different oceans. As the age of the last fifteen or so polarity changes is known (Figure 24, page 71) the rate at which each side of a ridge is moving away from its centre can be measured and is found to be equal in any one area, so that the opening rate is twice the spreading rate of the individual sides of the ridge.

Opening rate
cms./year

Ridge	Rate
East Pacific Rise	10–12
	8
Pacific–Antarctic Rise	6
S. Indian Ocean	
N. Pacific	
South Atlantic	3
N. Indian Ocean	2.5
North Atlantic	2

Millions of years

Distance from the ridge axis (kms)

240
160
80

1 2 3 4 5

Polarity sequence on the ocean floor
black = normal as today
white = reversed polarity

fracture zones, and they can be matched in sequences off-set on either side of the fracture zones. This striped magnetic pattern was attributed to filaments of the igneous floor being magnetized alternately in opposite directions and was clearly related to the changing polarity of the Earth's magnetic field, but it was still not known why this should result in such consistent linear patterns.

In 1963, the British oceanographers, Frederick Vine and Drummond Matthews, suggested the origin of this 'zebra' pattern (Figure 30). They proposed that igneous rocks were continually being intruded into the centre of the ridge system, where they cooled and became magnetized in the direction of the prevailing magnetic field. They were then gradually moved aside as new molten rocks were intruded along the ridge axis. These new rocks were magnetized in their turn and so a record of any changes in polarity of the Earth's field would be preserved in the rocks on either side of the ridge centre.

The confirmation of this suggestion came from a detailed survey of the magnetism of the Rekjanes Ridge, part of the Mid-Atlantic Ridge just south of Iceland. This survey was started in 1963 at the instigation of the American pioneer of many oceanographic studies, Harry Hess, and the analysis was completed in 1967. This showed that the sequence of polarity changes is exactly the same on either side of the ridge and that this sequence matches the table of reversals of the Earth's magnetic field established in 1964 (Figure 24, page 71).

This revelation led to the analysis of all previous oceanographic magnetic records and during 1968 and 1969 it was found that exactly the same reversal pattern occurs on all the ridges, but in some areas, such as the Pacific, the pattern is broader than in others. As we know the exact dates of the last few reversals, we can calculate

Magnetic anomalies (age in millions of years)

——— 0   ·········· 10   —·—·— 40   – – – 70

32. *Magnetic anomalies in the oceans*

This map shows some of the magnetic anomalies which were located by 1969. The study of these, and the detection of further anomalies in other parts of the oceans, is still being continued, but the overall pattern has already been established and has only changed in detail. (Anomalies in the North Atlantic are shown in slightly more detail in Figure 28, page 80.)

the rate at which the oceanic floors are moving away from their centres.

The North Atlantic is spreading at 1 centimetre a year in each direction, while each side of the Eastern Pacific is moving at 5 centimetres a year (Figure 31). Unfortunately we do not know the exact age of previous reversals of the Earth's field (as discussed in the previous chapter) but we can match the same sequence of polarity changes in different parts of all the oceans so that if we use the spreading rates calculated from the known reversal time scale, then the age of these other matching patterns can be estimated (Figure 32). In general, these older sequences are less than 80 million years old and in the Atlantic (Figure 28) and Indian oceans they extend very close to the continental slopes. This estimation of their age appears to be correct as it agrees with that of the sediments, and, as we will see in the next chapter, it also agrees with the rest of our evidence for the dating of the last stage of continental drift.

8 THE TIMING OF DRIFT

Much more precise data is required to date continental drift than merely to demonstrate that such movements have taken place, so our information is mainly restricted to areas for which we have the most geological knowledge, mainly the continents bordering the North Atlantic and other regions where detailed studies have been made. These are, of course, few at present as there has previously been so little support for the concept.

In the parts of the oceans where the magnetic pattern has been deciphered we can already time the separation of the continents during the last 75 million years or so, but elsewhere we must rely on the geological and palaeomagnetic information obtained from the continental rocks to time both the original fracturing of the great land masses and to estimate the rate of their later separation.

Three hundred million years ago, at the close of the Carboniferous and start of the Permian geological periods, the land masses of the world were grouped into two equal-sized, partly linked continents (Figure 33(a)) and the continents as we know them today were already distinguishable as parts of them. This grouping continued with very little change for the next 150 million years so that this is a convenient time to start our history, bearing in mind that this last 300 million years represents only one-fifteenth of the Earth's existence and many periods of drifting preceded this time.

Gondwanaland, comprising South America, Africa,

India, Australia with parts of S.E. Asia, and Antarctica, was positioned so that its southern part lay beneath the polar ice cap while its northern regions were in tropical latitudes. North America, Europe and Asia formed Laurasia which then straddled the equator.

Around this vast land mass lay long troughs in which sedimentary detritus was accumulating. These troughs later evolved into our modern circum-Pacific and Alpine-Himalayan mountains, but at that time and for most of their subsequent history, they were the sites of most of the Earth's geological activity as their oceanic sides, in particular, were marked by numerous volcanoes. (The geography of these troughs was remarkably similar to our deep oceanic trenches discussed in the next chapter.)

Between Laurasia and Gondwanaland lay the Tethys Ocean. This narrowed westwards so that these super-continents were linked from north-west Africa to North America and southern Europe, although shallow seas often covered this link preventing the migration of terrestrial plants and animals.

Most of Gondwanaland was dry land at that time, forming a huge basin-like area. This was being eroded by the wind and rivers which carried the debris into the lower, central parts of the basin or over the edge of the continent into the bordering troughs. Along the southern side of the Tethys, shallow seas covered parts of north Africa, India, and north-eastern Australia but these receded from the land during the next 50 million years as did the shallow seas which existed in the Amazon Basin. Many fractures were already present which were critical to the later formation of both the South Atlantic and Indian oceans. There is, in fact, some evidence that a proto-Indian Ocean, a small sea similar to the present Mediterranean, existed at this period near western Australia—the *Sinus Australis*. The western edge of

Australia had been an actively subsiding area for the previous 100 or even 200 million years, but as it subsided it was filled with sediments worn off the Australian continent and only rarely sank low enough for marine waters to cover it. A similar situation existed along the East African margin where 7 to 8 kilometres of terrestrial sediments accumulated in a subsiding trough. There is no evidence that this trough was below sea level except for very brief periods such as 250 million years ago, when seas stretched from the west coast of India to Madagascar.

During the next 150 million years (Figure 33(b)), Gondwanaland changed very little except that it drifted very gradually northwards. (This movement seems to have been accompanied by a general amelioration of the world's climate until the northern continents finally enclosed the north polar regions some 3 million years ago, but we will be discussing the relationship between Ice Ages and continental drift in Chapter 10.) Towards the end of this period, this vast basin gradually became a series of separate basins although the fundamental picture of erosion and deposition still continued.

---

33. The distribution of the continents (overleaf)

The four diagrams attempt to illustrate the relationship of the continents at four points in time, but the presence or absence of shallow seas only reflects an average picture at each time. The latitudes and longitudes have been omitted (the projection is stereographic, centred on Africa) as our evidence for their exact location is not sufficiently accurate at present, although Gondwanaland in general was drifting from polar regions to equatorial (compare with Figure 18, page 57), while Laurasia was gradually drifting northwards. The heavy lines mark the position of the fractures which opened to form the Indian and Atlantic Oceans and these are shown broken while the fracture is developing but has not fully separated. No allowance has been made for minor movements within the continents, such as the uncurling of the Patagonian 'tail' of South America from around Africa.

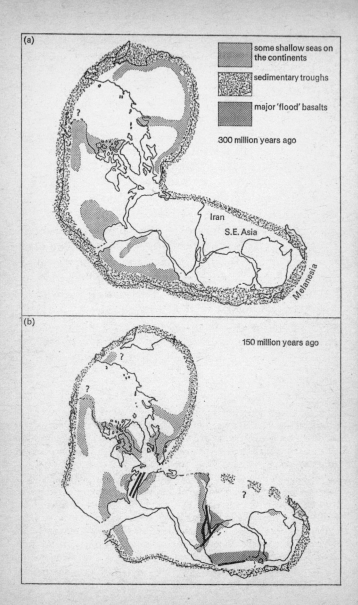

(a)

some shallow seas on
the continents

sedimentary troughs

major 'flood' basalts

300 million years ago

Iran

S.E. Asia

Melanesia

(b)

150 million years ago

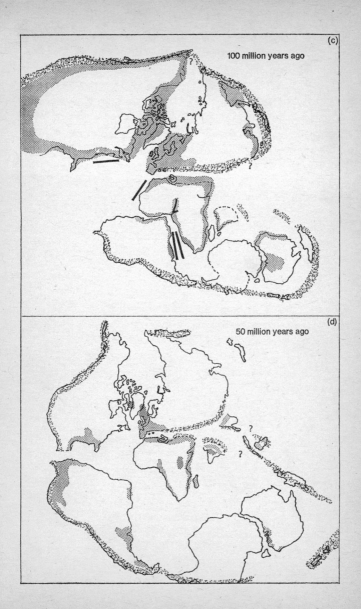

(c) 100 million years ago

(d) 50 million years ago

Suddenly there was a violent disruption of the scene as vast floods of basaltic lavas spread out on all the southern continents at almost exactly the same time, covering many thousands of square kilometres. Indeed the size of these floods indicates that Gondwanaland must have been floating on a cushion of molten rocks which were released as the existing fractures opened slightly and linked to form the cracks which divided Gondwanaland into our present continents. These lavas flowed about 160 million years ago in Australia (Tasmanian Dolerite), India (Rajmahal Traps), the Middle East (Yemen Traps), Antarctica (Ferrar Dolerites), and southern Africa (Karroo Lavas). The only Gondwanan continent which escaped this initial volcanic activity was South America, where similar lava flows (Serra Geral) did not appear for another 40 million years when they were accompanied by further eruptions in southern Africa.

The release of the lavas allowed the areas near the cracks to sink, forming basins in which water collected and evaporated, leaving behind accumulations of salt which can now be found bordering most of the southern continents. As the areas sunk even lower, narrow strips of sea water spread from the Tethys Ocean along the edges of the continents, forming the predecessors of our present deep oceans.

These narrow, shallow seas, resembling the present Red Sea (Plate IV) and Gulf of California, can be traced as they spread southwards, passing between India and Somaliland some 150 million years ago, to link with seas already existing near Madagascar and South Africa. About 10 million years later, similar seas spread along the east coast of India. Australia, at that time, was almost covered by vast shallow, freshwater lakes. Unfortunately most of the evidence for these marginal seas now lies on the continental shelves, beneath the oceans, and we must rely on the isolated patches of marine sediments left behind on what is now the dry land of the continental edge. The split between Australia and Antarctica occurred late in the history of the break-up of Gondwanaland, the separation starting about 50 million years ago. The flood basalts on Antarctica and Australia, which are 160 million years old, seem to be related to the movement of Western Antarctica relative to Eastern Antarctica as the separation of New Zealand from Antarctica–Australia was also fairly late – approximately 80 million years ago. The evolution of South East Asia is even more complex; there is evidence for movements between several little 'micro-continents' during much of this time – some movements welding fragments together and some splitting blocks apart. These areas, however, are geologically complex and difficult to investigate, so that even the present situation is not very clear.

On the western side of Africa and along eastern South America there is much more evidence and the detailed way in which the marginal seas progressed along the present coastlines has been described by the British stratigrapher, Richard Reyment. The seas reached South West Africa 120 million years ago, the Congo some 10 million years later and Nigeria 105 million years ago. This progress was made in a series of advances and retreats, each advance carrying the seas farther northwards until water existed almost continuously between Africa and South America by 100 million years ago (Figure 33(c)). At that time the seas were filling a fracture, the Benue Trough, which cuts across the bulge of Africa from Nigeria into Algeria, and almost connects to the Mediterranean. However, instead of driving farther along this fracture, the next advance took the seas westwards to link with marginal seas which had been slowly making their way eastwards between the bulge of Africa and northern

Brazil, so that 92 million years ago, the marginal seas merged, finally separating Africa from South America.

The subsequent history of the Indian and South Atlantic oceans is revealed by the palaeomagnetism of their floors and of the rocks on the continents. Although the opening of the Indian Ocean commenced some 160 million years ago and the South Atlantic some 120 million years ago they did not widen very much until about 100 million years ago. Between 100 and 80 million years ago, South America rotated away from Africa before it drifted westwards to its present position, and meanwhile India rotated away from Africa and moved northwards until it eventually came into collision with Asia. Australia and Antarctica appear to have drifted away from Africa together before finally separating into their present positions during the last 50–60 million years, while Africa rotated slightly and moved northwards to impinge on Europe.

Laurasia, at the start of this last 300 million year period, was generally lower lying than Gondwanaland, except for the Appalachian-Caledonian mountains (Chapter 3), the Russian Urals, and the eastern Siberian mountains which were all high and were only gradually worn down during the next 50 million years. Most of North America was covered by seas, similar to Hudson Bay today, which slowly receded to the south and west during the next 150 million years. Europe, west of the Urals, was also covered by shallow seas spreading northwards from the Tethys which persisted, although shifting their location from time to time, for much of the following 200 million years. In the Far East, marine sedimentary troughs occupied much of the north-east Asian region, which was isolated from China and parts of S.E. Asia at the beginning of this period. From these troughs, shallow seas spread out over parts of China and western Siberia,

1 VOLCANOES —SURTSEY AND KAO

The upper photograph shows the new island of Surtsey which formed off southern Iceland in a series of spectacular eruptions commencing in 1963. This volcano is one of the many volcanoes which have formed as the continents have drifted apart and new ocean floor has been added at the middle of the oceans (page 86). The lower photograph is of the island of Kao in the Tongan Islands. This volcano is quite different from Surtsey, having formed near the deep-sea Tonga Trench where old ocean floor is returning to the Earth's interior (page 112).

The origin of both types of volcano can now be understood from our new view of continental drift and it is becoming possible to predict their future activity and whether this is likely to be dangerous and highly explosive, as in Kao, or comparatively quiet and only dangerous at the actual point of eruption, as in Surtsey (page 122). (Upper photo: Icelandic Survey Department; lower photo: D. H. Tarling)

2 THE RIFT VALLEY IN ICELAND

Iceland is one of the few areas in the world where the valley which follows the mid-oceanic ridge (page 74) can be seen on land. This photograph shows the rift, the Thingvellir Graben, at Almannagjá in the south-west of the island, where the two sides of the Atlantic ocean floor are separating as Europe and North America drift apart. As the separation takes place, molten rocks from the Earth's interior are able to reach the surface, filling the gap and building new ocean floor (Figure 30) as they solidify. (Photo: S. Thorarinsson)

3 THE SAN ANDREAS FAULT IN CALIFORNIA

The San Andreas Fault, seen here running through the Elkhorn Mountains in the Great Valley of California, is the boundary between the two plates of the Earth's surface (Figure 41). The Earth's crust on the right of the photograph is moving in the opposite direction to that on the left. We know that this movement should average 5 centimetres every year, the left-hand side moving away from the reader. If this movement does not take place, then stresses accumulate until the friction is overcome and the two sides suddenly readjust to their proper position, causing a major earthquake (page 118). With this knowledge it should be possible soon to use new techniques to prevent major earthquakes occurring in this region. (R. C. Frampton and J. S. Shelton photo)

4 THE NORTHERN RED SEA AND SINAI

The Red Sea, seen here from a satellite, has formed during the last 5 million years as Arabia has separated from the African continent, and new oceanic crust has formed in between. In much the same way, the Baja California has also separated from the American mainland as new oceanic material has formed along the Gulf of California. These areas are in the earliest stages of ocean formation (Figure 43) and are similar to the North Atlantic some 150 million years ago. The formation of these early rifts seems to be associated with the emplacement of many valuable ore deposits so that an understanding of oceanic evolution may have direct economic significance (page 126). (A NASA photograph)

although most of Asia remained dry land throughout its subsequent history.

The first strong geological activity, apart from that along the marginal troughs, occurred in Siberia, 200 million years ago, when flood basalts, up to $2\frac{1}{2}$ kilometres thick, spread out over 500,000 square kilometres. These eruptions almost certainly relate to some major movement, which re-elevated the Urals, originally formed some 300 million years ago but which had been subsequently eroded (they were then eroded yet again and then elevated yet again during the last 20 million years). This movement was probably related to movements in S.E. Asia, but the evidence is too sparse at the moment for a realistic evaluation.

The main framework of fractures destined to link and open to form the North Atlantic was already in place 300 million years ago and had during the preceding 50 million years controlled the location of rich metalliferous ores, mainly silver and zinc, which now occur in Newfoundland and the British Isles. However, separation commenced further south, for although the northern continents were linked to Gondwanaland at the start of this period, North America broke away from Africa 180 million years ago. This break, forming the proto-Central Atlantic, accompanied the eruption of some lavas in both eastern North America and Morocco, although not on the scale of the vast eruptions in Siberia or the southern continents. This movement left Europe still connected to North America, but formed a shallow sea, possibly 300 or 400 kilometres wide, in which a thickness of 3 or 4 kilometres of sediments accumulated which can now be traced along the eastern edge of the Atlantic continental shelf of North America. This initial opening does not seem to have been particularly violent or rapid and may even have ceased before the onset of the next phase of movement when the Atlantic began to widen to approximately a quarter of its present width. This stage gave way some 70–80 million years ago to the present more rapid expansion of the Atlantic. The Labrador Sea lies on a very old fracture zone which appears to have widened between about 50 and 70 million years ago, but then stopped at approximately the same time as the northern part of the North Atlantic, between Greenland and Europe, began to open, 55 million years ago; this opening being preceded by the eruption of flood basalts in northern Britain and East Greenland.

We have, therefore, quite a complex history of the opening of the present oceans, but there were several features in common. In particular most of the fractures were present long before they cracked apart sufficiently to allow the escape of lavas, and subsequent to these initial openings, between 150 and 100 million years ago, the rate of separation was generally slow as about two-thirds of the Earth's present ocean floors have formed during only the last 80 million years.

It is much more difficult to detect common features in the development of the sedimentary troughs bordering the continents. Most of these have a long and disturbed history; different areas underwent compression and intrusion by igneous rocks at different times. This activity was well developed at the start of this last 300 million year period and has continued for most of their subsequent history. However, during the last 20 million years all these troughs seem to have been similarly affected but it was mainly during the last 20 million years that they underwent their most intense phase of deformation when the majority were converted into the mountains we see today. The explanation of this major disturbance of the troughs of the Tethys is easy to explain as at that time they were being crushed between the northward-moving southern continents, Africa and India, and the slower-moving northern continents, Europe and Asia. Much of the compression of the circum-Pacific troughs is clearly related to the movement of the continents which carried the troughs with them during the last 70–120 million years, but it is not clear, at present, why they did not undergo their most intense compression and uplift until recently. It may relate to some change in the forces causing continental drift during the last 20 million years (discussed further in Chapter 10).

We have, then, quite a reasonable picture of the general evolution of our drifting continents for much of the last 300 million years. For any earlier study we have to rely on palaeomagnetic and geological arguments and, at present, there is, unfortunately, little palaeomagnetic data available for the older rocks of the continents; in any case these old rocks are difficult to date accurately, particularly as fossils can only be used for this purpose during the last 570 million years (Chapter 3).

However, we do have good evidence for continental drifting some 400 to 450 million years ago. Identical shallow-sea fossils some 500 million years old can be found in Scotland and northern Newfoundland and these are quite distinct from other shallow-sea fossils of the same age which can be found in England and southern Newfoundland, yet there is no evidence of a land barrier having existed between these northern and southern areas. The British geologist, John Dewey, has suggested that these fossils once lived on the opposite shores of an ancient ocean and that the continents bordering this ocean drifted together some 450 million years ago, squeezing the oceanic sediments into the Appalachian-Caledonian mountains and bringing within 50 kilometres of each other, organisms which had died thousands of kilometres apart.

We have been looking mainly at the latest phase of movements of continental masses, but there is evidence for similar movements throughout most of the Earth's history, and certainly for the last 2,700 million years. Consequently there have been many previous patterns of land and ocean upon the Earth. We must, therefore, consider the nature of the forces that are continually changing the Earth's surface.

The forces causing such tremendous movements of the surface of the Earth must be situated in its interior; it is inconceivable that an external force could cause such persistent movements in opposite directions on a spinning and orbiting planet. Until recently most people doubted the existence of such internal forces, but our improved knowledge of the Earth's interior has shown that adequate forces could exist and, with the acceptance of the theory of continental drift following the new evidence from the oceans, we can now study how these forces operate.

Most of our knowledge of the interior of the Earth comes from the study of sound waves. When a large earthquake or an atomic explosion occurs, powerful sound (seismic) waves radiating out from it can be detected and measured after they have travelled through the Earth, giving an X-ray type picture of its structure. There are two types of these seismic waves, surface and body waves, and their speed and energy is controlled by the physical state and composition of the rocks through which they pass.

The surface waves, which travel much more slowly than the body waves, are so called because their crests always travel at the surface of the Earth, but they give us detailed information of the deep structure of the Earth because the depth of penetration of the waves is determined by their wavelength. By examining the short

wavelength surface waves we can study the physical properties of the top surface of the Earth and using longer
wavelengths we can determine the average properties
down to greater and greater depths and so build up a
composite model of the Earth. There are two types of
body waves which travel out in all directions through the
Earth. The faster P waves can pass through both solids
and liquids but the slower S waves can only pass through
solids. By measuring the different magnitudes and arrival
times of these body waves at seismic stations around the
world, a picture of the nature of the Earth's interior was
developed in the early twentieth century, and this has
recently been modified in detail, and for our understanding of continental drift, very significant detail, by observations using instruments and techniques developed for
monitoring nuclear weapon testing and from the previously neglected study of surface waves.

The Earth (Figure 34) contains a core with a radius of
3,490 kilometres, just over half of the Earth's 6,415 kilometre radius. The S waves are unable to pass through the
outer part of this core, showing it to be in a liquid state –

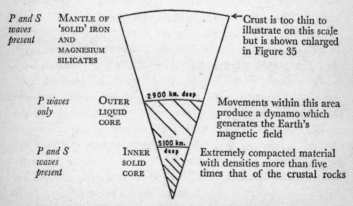

P and S waves present	MANTLE OF 'SOLID' IRON AND MAGNESIUM SILICATES	←Crust is too thin to illustrate on this scale but is shown enlarged in Figure 35
P waves only	OUTER LIQUID CORE	2900 km. deep — Movements within this area produce a dynamo which generates the Earth's magnetic field
P and S waves present	INNER SOLID CORE	5100 km. deep — Extremely compacted material with densities more than five times that of the crustal rocks

34. *A segment of the Earth as revealed by seismic waves*

that is liquid in the physicist's sense, i.e. without rigidity, but as the material is at a high temperature and under such tremendous pressure, well over one million times the Earth's atmosphere, it does not behave in exactly the same way as a liquid at the Earth's surface. Within this liquid outer core is a small, very dense solid inner core.

Overlying the core and forming the major part of the Earth is the solid mantle, composed of material similar to the densest rocks found at the surface of the Earth. It was originally thought that these mantle rocks gradually increased in density with depth as they are compacted under greater and greater pressures, but we now know that most of the increase takes place as a series of distinct jumps. Each of these density discontinuities marks the depth at which the weight of the overlying rocks forces the atoms in a specific mineral structure to re-adjust their positions to a more compact crystal form. It is not easy to study these density changes in the laboratory as it is difficult to reproduce the high pressures and temperatures existing within the mantle. We can only simulate, for a practical length of time, the physical conditions which occur at a depth of about 150 kilometres (although it is possible to produce pressures higher than those at the centre of the Earth for a split second during the impact of a bullet on a target). From these high pressure studies, the Australian geochemist, Edward Ringwood, has been able to predict the depths at which these phase changes take place in minerals which form much of the mantle. For example, olivine, an iron-magnesium silicate, the main constituent of mantle rocks, will change from its normal surface structure to a higher density garnet form at a depth of 300 kilometres, then at a depth of 400 kilometres will collapse to an even denser spinel form which again collapses to a still more compact structure at a depth of about 800 kilometres.

The Earth's mantle is covered by the crust (Figure 35) which varies in both thickness and composition. For the continents it is generally some 20 kilometres thick, but is somewhat thicker beneath old, worn-down mountains and is much thicker, up to 50 kilometres, beneath modern mountain chains. As we have seen, the composition of the rocks forming the continental crust is very variable. They have an average density of 2·67 near the surface (density being measured by comparison with water of density 1·0) but their density increases to 2·77 near the base of the crust. The boundary between the continental rocks and the high density, 3·3, rocks of the upper mantle is very sharp and is called the Mohorovičić Discontinuity (Moho) after A. Mohorovičić who discovered it in 1909 when examining the seismic records of an earthquake in Yugoslavia (Croatia).

The Moho can also be located beneath the ocean basins where mantle rocks of the same density are in sharp contrast to the rocks of the oceanic crust which have a density of 2·9. The ocean crust has a uniform thickness of about 8 kilometres and is formed of three

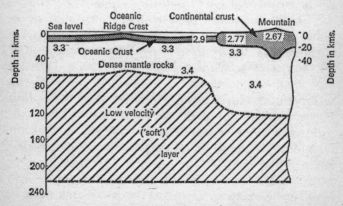

35. The crust and upper mantle

distinct layers. The upper layer comprises the sediments which we discussed in Chapter 7. These overlie the upper igneous volcanic layer, the top $\frac{1}{2}$ kilometre of which carries the magnetic pattern used to date the ocean floor. Denser igneous rocks form the main oceanic layer some four to five kilometres thick.

There are, therefore, distinct differences between the oceanic and continental crusts, some of which were used to define a continent in Chapter 2. One of these differences is that the continental crust thickens where the land is higher, but the oceanic crust remains almost constant in thickness over the ridges although these stand 4 kilometres high above the normal ocean floor. It was originally thought that this implied that there was more crustal material beneath the ridges than under the rest of the ocean basins, but studies with gravimeters (Chapter 2) showed that this was not so; the amount of rock (total mass) beneath the oceanic ridges is the same as beneath the rest of the ocean floor. This means that the mantle rocks beneath the ridges must be of lower density and, therefore, take up more room. This interpretation of the gravity observations was subsequently shown to be correct by seismic investigations across the ridge and studies using both of these geophysical methods have now shown how this low density material thins away from the crest of the ridge (Figure 36).

As long ago as 1926, the American seismologist, Beno Gutenberg, suspected the presence of large scale, low-density material within the mantle, but the confirmation of its existence has only come within the last decade from the study of body waves from atomic test sites and surface waves from earthquakes. With controlled atomic explosions we do not have the problem of estimating their exact time and location as we do for earthquakes. However, it was mainly from the examination of the 1960

Chile and 1964 Alaskan earthquakes that an extensive low density layer was mapped within the mantle. These studies showed that this 'soft' layer (Figure 35) lies in the upper part of the mantle; starting about 60 kilometres beneath the oceans and 120 kilometres beneath the continents and reaching to a depth ranging from about 200 to 250 kilometres. This layer is obviously of major importance and studies are still in progress to establish its detailed distribution and properties.

The reason for the existence of this low-density, 'soft' layer appears to be straightforward as the rocks at this depth are almost molten and, therefore, less dense. The interior of the Earth is clearly hotter than its surface, as shown by volcanoes and the temperatures within mine shafts and drill holes. Measurements taken underground

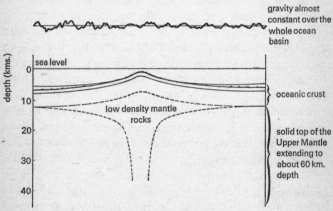

36. A section through a mid-oceanic ridge

The general smoothness of the gravity field over the ridge shows that the height of the ridge above the surrounding ocean floor must be compensated by low density rocks within the mantle. The presence of these rocks has subsequently been traced using seismic waves to penetrate the crust and reflect back to detectors at the ocean surface.

in the geologically stable parts of the world, such as the South African gold mines and European coal mines, show a temperature increase of 17°C per kilometre and this rate of increase probably continues down as far as the 'soft' layer where the mantle rocks are hot enough to be nearly molten. However, this rate of temperature increase cannot persist at greater depths as the mantle rocks would become completely molten, yet the seismic observations show that they become increasingly solid beneath the 'soft' layer.

Although the amount of heat escaping from the interior of the Earth has been measured on land for many decades, it was not until 1954, when Sir Edward Bullard developed an instrument for measuring the rate of heat flow through oceanic sediments, that world-wide studies could be made. The instrument is basically a standard core barrel used for sampling deep oceanic sediments (Chapter 7), but modified to measure their temperature at different levels. The measurements soon showed that there were regional variations in the rate of heat flow from the mantle, with high rates near the oceanic ridges and low near the deep ocean trenches (Figure 37).

Similar horizontal differences in density within the mantle have recently been inferred when using satellites to measure the exact shape of the Earth. These measurements of the Earth can only be explained by regional variations of density within the mantle. As there are variations of both density and temperature within the mantle it seems very likely that there is a circulation of hot, less dense material rising towards the surface, spreading out and cooling, then sinking as dense material. This circulation would carry the upper solid skin of the mantle and the crust of the Earth (i.e. the rocks overlying the 'soft' layer) from the hot, rising areas towards the cooler, sinking areas, thereby forming a system of convection

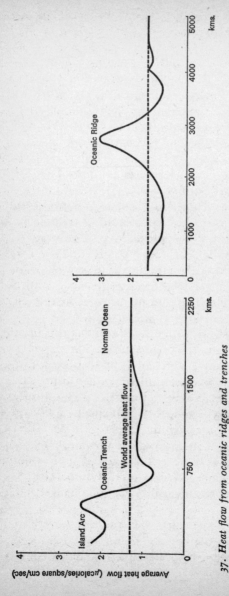

37. *Heat flow from oceanic ridges and trenches*

The graphs show the average values of heat flow; the individual observations show a very wide scatter over the ridge centre, some being ten or twenty times the world average, but over most of the ocean floor the heat flow is uniform.

currents similar to those suggested in 1928 by Arthur Holmes.

One of the main obstacles to the acceptance of the presence of convection currents was the seismic evidence for a solid mantle, but when we consider the behaviour of solids under very long term forces we find that they do, in fact, possess a plasticity (rheidity). For example ice is certainly solid and yet glaciers not only slide downhill but actually flow; similarly rock salt flows under pressures sustained for tens of years and even granite flows if the forces persist for thousands of years. Therefore, seismic waves, which pass through the Earth in minutes (about 20 minutes from the earthquake to the opposite point on the surface of the Earth) can give no indication of the long term properties of the mantle rocks for they would appear to be solid to the passage of seismic waves over a few seconds but behave as a liquid under forces maintained for millions of years. The ability of at least some mantle rocks to flow under sustained pressure can be seen in many examples within the geological record.

During the last ten thousand years, a huge ice cap which once covered Scandinavia has melted. While it was present, it weighed down that part of the continental crust and since its disappearance, the whole of Scandinavia has been rising, and, in fact, still has another 100 metres to rise to regain its former height in the centre of Sweden. This movement of the crust shows that the material of the mantle must be able to flow away when the crust is loaded and to flow back when the load is removed. Other isostatic examples are plentiful – the area around Salt Lake City has risen as a once huge lake, Lake Bonneville, evaporated, leaving behind the comparatively small Great Salt Lake. On a larger scale, along the western edge of Australia, as we have already mentioned, a slice of crust has been sinking continuously for at least 400 million years simultaneously filling with sediments now 7 to 8 kilometres thick. Conversely, the Colorado Plateau of the western United States has risen by 2 kilometres during the last few million years and mantle rocks must have entered this region at the same rate as the rise. Man himself has also produced some flow of mantle material; for example the water now held behind the Boulder Dam is heavy enough to depress the crust of the Earth slightly. In fact it is just possible to discern the effects of the ocean tides in the flexing of the crust of the continental shelves.

It is evident that the mantle rocks, being capable of flow, would move under the influence of temperature and density differences and the most recent estimate suggests that flow from hot regions to cold takes place when a very small temperature difference persists. So there can no longer be any question of the existence of convection currents in the upper mantle. Our new knowledge of the ocean floors shows that these currents carry the hot mantle material up into the oceanic ridge centres and then out in opposite directions away from the ridge. This new crustal material can only be accepted if the Earth is expanding, if its surface is buckling, or if other surface material is returning into the interior.

The idea of an expanding Earth has had a long and interesting history, but recent observations have shown that any expansion is unlikely or so extremely small that it is not significant during at least the last 1,000 million years. This was verified by counting the number of days in the year, and in the month, during the geological past. These can be calculated for 400 million years ago by studying the growth rings on Devonian corals. These have daily growth rings in a similar way as trees develop annual rings, though the coral daily growth rings are superimposed on monthly and yearly growth bands.

Counts of these rings show that there were 400 days in the year at that time, which means that the Earth must then have been spinning faster but nowhere near as fast as it would have been if it was much smaller than today. This rate of spin is, in fact, what would be expected at that time as we know the Earth is being very gradually slowed by the action of tides on the continental shelves. Similarly palaeomagnetic observations can be used to measure the ancient radius of the Earth and they do not suggest any significant change during the last 400 million years or so.

The only indication of major buckling of the surface of the earth is within its mountain belts. These show clear evidence that the rocks in them have been compressed and many attempts have been made to unravel the contorted rock layers in order to estimate the amount of compression. This has led to extremely variable figures because many of the surface features of mountain belts result from the very gradual bending over of great folds of rocks which are further contorted as they 'flow' downhill under the influence of gravity (this plastic flow is similar to that within the mantle). However, most estimates for different mountain systems, for instance the Alps or the Appalachians, suggest that their compression is of the order of 200 to 250 kilometres. This scale of compression in all our modern mountains would amount to very much less than the amount of crust which has been added to the ocean floor in the last 25 million years.

As we have seen, our present oceanic crust is less than 200 million years old so that all the previous oceanic crust, which for most of the Earth's history must have formed two-thirds of the Earth's surface, must have been absorbed at the sites of downward convection currents. The most recent studies of the Earth show that this ingestion is taking place today in the deep oceanic trenches of the world. These trenches (Figure 25, page 75) are the deepest parts of the Earth's surface, usually around 8 to 10 kilometres deep (the deepest point being in the Marianas Trench in the West Pacific, which is just over 11 kilometres deep). The trenches are of obvious geological importance as they are all related to strong earthquake and volcanic activity. The continental side of the trenches are usually marked by island arcs formed by active volcanoes (Figure 42, page 123) which overlie almost all of the world's deep earthquakes. Both the trenches and the island arcs show many characteristic features, in particular major gravity and magnetic anomalies. However, it is rather difficult to generalize about their more detailed features. The trenches are usually V-shaped in section, but some are filled with sediments. Although they are only tens of kilometres wide, they are hundreds of kilometres long, and vary in shape from straight lines (Kermadec and Tonga) to angular (Solomons) and from smooth curves (Aleutians, Marianas) to complex arcs and twists (Banda Sea, Celebes). In some areas these shapes seem to reflect the shape of the continent which they border.

The main clue to their geological significance lies in detailed studies of the positions of the earthquakes (Figure 38). These are shallow beneath the trenches, scattered from the surface down to about 80 kilometres, but at greater depths they are situated correspondingly closer to the continents. In the regions where adequately detailed studies have been made these deeper earthquakes all occur within a narrow band some 15 to 20 kilometres thick which dips down from the trench at an angle of approximately 45°, although tending to become steeper at greater depths. These earthquakes, therefore, map the surface of the oceanic crust and upper mantle rocks as they are carried down into the interior of the

Earth. As these rocks begin to descend they are bent, their cracking giving rise to the shallow earthquakes, but at greater depths the nature of the earthquakes differs and here they probably result from the physical and chemical changes of the rocks as they are carried into regions of greater pressure and temperature. It is possible, in fact, that the crustal rocks are converted into denser mantle rocks and sink, thereby assisting the downward convection. Below some 700 kilometres these crustal rocks must merge into the mantle rocks as no earthquakes usually occur below this depth. (The deepest known earthquake occurred in 1924 beneath the Celebes at a depth of 720 kilometres.)

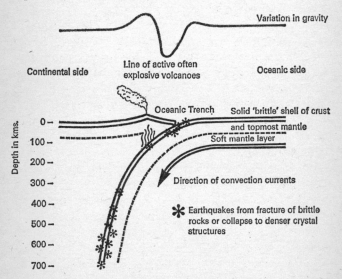

38. The relationship of earthquakes and trenches

The oceanic crust is carried down into the mantle, together with the brittle part of the mantle overlying the 'soft' layer, and this gives rise to the intense earthquake and volcanic activity of the trenches and their marginal island arcs.

When the crustal material is carried into the mantle, it has abnormally low density compared with mantle rocks, accounting for the variation in the Earth's gravity, and as it becomes hotter, water and other volatiles are driven off and these form a gas-rock mixture which rises to the surface forming volcanoes, usually explosive, such as those rimming the Pacific.

Many of these trenches have still to be studied in

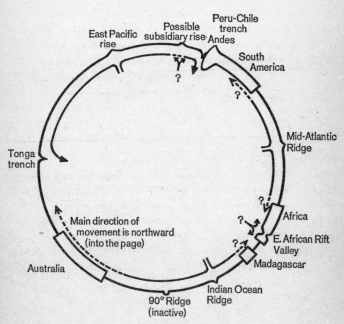

39. *A slice through the Earth at 20°S latitude*

The relationship of the oceanic ridges and trenches to the uprising and down-going convection currents is well established, but we know little concerning their action beneath the continents and their relative movement (marked by question marks and broken lines), while the extension of these convection currents into the deeper interior of the Earth is still a major problem.

detail, but there seems to be good evidence that the older parts of the Pacific crust are being 'digested' at a rate of about 12 centimetres per year in the Japanese Trench and at even faster rates in the Aleutian Trench. It is certain that the remaining trenches of the world are also able to 'digest' material so that the addition of newly formed crust to the Earth's surface does not increase its size.

The major problem which faces us at the moment is the extent of these convection currents. Although they rise at the oceanic ridges and sink at the trenches (Figure 39), with their upper surface at the top of the 'soft' layer, it is by no means clear if these currents circulate only in the top 300–700 km. or if they travel throughout the depth of the mantle. It seems physically difficult for rocks to circulate beneath the 'soft' layer where minerals must change their structure at the density discontinuities, but there are both physical and chemical indications that these currents, at least at some stage in the Earth's history, circulated throughout the mantle. This, however, is only important for more detailed studies of continental drift and we can now look at the Earth to see how this picture of continental drift and convection currents helps us to understand its past history, its present surface patterns, and future evolution.

The drifting of the continents and ocean floors over the surface of the Earth is not merely an interesting, purely academic, observation, for the study of these movements is leading us to the understanding of earthquakes and volcanoes and the way in which our mineral resources are distributed.

Most earthquakes occur at the sites of downward convection currents, but rising currents produce a series of weak, shallow earthquakes in the central valley of the oceanic ridges where new crust is being added, and so a map of the distribution of all earthquakes (Figure 40(a), (b)) is essentially a map of the action of convection currents. This completely new type of map shows that the earthquakes outline vast slabs of Earth, some 80 to 100 kilometres thick, which are moving individually over the 'soft' layer of the mantle (Figure 40(c)). As each slab is moving at a constant rate in a particular direction there is very little geological activity within it and the world's earthquakes and volcanoes are concentrated along the margins where the slabs or plates interact. The study of the surface movements of our present-day Earth is now termed 'plate tectonics' and has evolved from 'continental drift'.

With our new understanding of the origin of earthquakes it is becoming possible to predict their occurrence and magnitude, and ultimately to control them. In California, for example, an area becoming increasingly

40(a). The distribution of earthquakes and tectonic plates

Most earthquakes occur in narrow bands, such as those occurring between January 1965 and December 1967 illustrated above and overleaf (a & b). Bands can be used to outline plates of the Earth's crust (c). Each of these plates move uniformly and independently so that the Earth's geological activity is mainly restricted to the areas where these plates are interacting. (Figures a & b include earthquakes produced by nuclear weapon testing.)

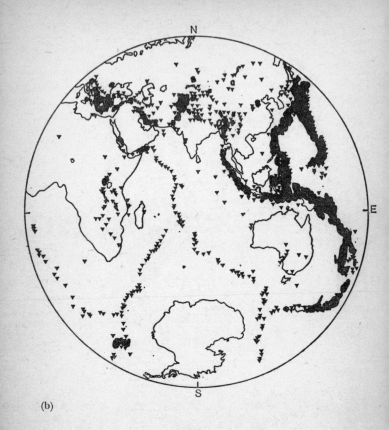

(b)

densely populated, earthquakes occur along a series of fractures in the Earth's crust, the San Andreas Fault System. In 1906, movement between the two sides of this fault system destroyed most of San Francisco and a similar earthquake in this region would still be a major disaster despite building precautions and improved engineering techniques. (These techniques would have saved many thousands of lives in Morocco when Agadir was destroyed in 1960 but could not save lives in California

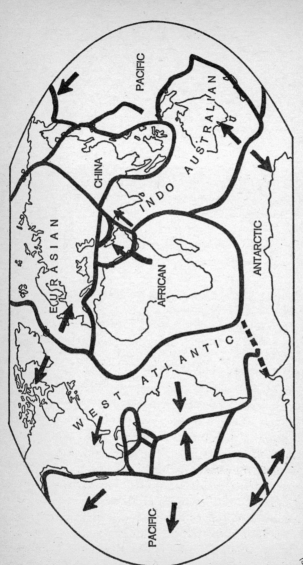

PACIFIC

PACIFIC

CHINA

INDO AUSTRALIAN

EURASIAN

AFRICAN

ANTARCTIC

WEST ATLANTIC

©

should an earthquake occur there as severe as that which, in 1811–12, devastated over 12,000 square kilometres around New Madrid, Missouri.)

On the map of tectonic plates (Figure 40(c)), the San Andreas fault system (Plate III) can be seen to lie on the boundary between the Pacific and West Atlantic plates at a point where the Pacific plate is moving north-north-west relative to the Atlantic plate by some 5 centimetres a year (Figure 41). Although very small and harmless movements do occur along the fault system, its sides in general are held firm by friction which is not overcome until sufficient strain has accumulated to cause the fault to give way resulting in a major earthquake as the two sides adjust themselves to their correct position. In populated earthquake areas measurement of these small movements is continually going on, and, now that the average annual movement which should take place is known from plate tectonic studies, the amount of accumulated strain can be calculated and the magnitude of an earthquake in the near or distant future can be predicted. These calculations will enable the authorities to make the best possible arrangements to minimize the disaster but unfortunately does not allow the date of the earthquake to be determined, although the probability of an earthquake occurring clearly increases as the amount of accumulated strain increases.

In the very near future it should be possible to control some earthquakes. For example, fault systems, such as the San Andreas, could be lubricated by pumping water into them at depth – this would lessen the friction points and movement would occur as a series of small shakes and not build up into a major earthquake. Alternatively, small relieving earthquakes could be triggered off by appropriately placed large, probably nuclear, explosions. These methods could only be effective because of our new

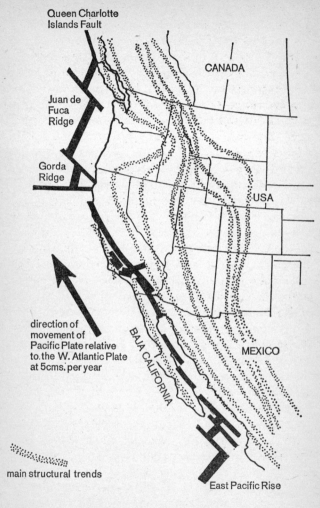

Queen Charlotte
Islands Fault

CANADA

Juan de
Fuca
Ridge

Gorda
Ridge

USA

direction of
movement of
Pacific Plate relative
to the W. Atlantic Plate
at 5cms. per year

BAJA CALIFORNIA

MEXICO

main structural trends

East Pacific Rise

41. Continental drift and earthquakes in California

knowledge of plate tectonics depending as they must on an accurate assessment of the amount of movement necessary to relieve any accumulated strain.

Tidal waves (more properly called *tsunami* as they are not related to tides controlled by the Moon) appear to be mainly generated by earthquakes at trenches where the plates are colliding head on, so that one plate is being thrust directly beneath another. This situation arises particularly in the largely uninhabited Aleutian arc, off Alaska, and suggests that this area should be most closely monitored to give maximum warning to the rest of the Pacific of the onset of tsunami. As tsunami are related to earthquakes, there is clearly the possibility that these can be predicted in this and other areas in the same way as earthquakes.

The volcanoes of the world are also related to the plates and their movements. At the moment we can only map (Figure 42) the distribution of active volcanoes on land or where they break the surface of the oceans, yet as all the ocean floors have been formed by volcanic rock intruded at the centre of the oceanic ridges, the vast majority of the world's present (and past) volcanic activity must be occurring at these ridges.

Geologists have long recognized two main types of volcano (Plate I). The one type produces the quieter *basaltic* eruptions such as those on islands in the deep oceans (these are safe enough to provide tourist attractions in such places as Hawaii and Iceland), while the other type produces the usually violent *andesitic* erup-

42. *The distribution of active and recently active volcanoes* (opposite)
Most of the volcanoes of the Earth occur beneath the ocean waters and cannot be mapped at present, but the volcanoes occurring on land are mainly found along the edges of continental plates where the oceanic plates are descending.

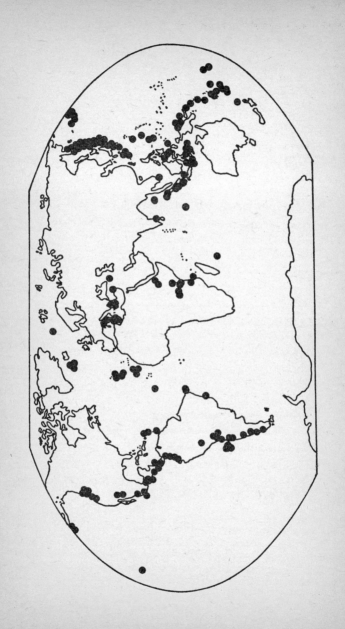

tions which mostly occur on the continental side of the oceanic trenches. As we have seen (Chapter 9), these andesitic volcanoes arise from the gases and liquids escaping from the oceanic crust as it is carried down into hotter regions of the mantle and their detailed composition depends on the depths at which these volatiles are driven out (Figure 38, page 113). This means that the detailed composition of andesitic volcanoes can be predicted when the depth of the upper surface of the descending crust is known and this can be determined from the depth of earthquakes immediately below the volcanoes. Conversely, the composition of ancient andesites of the world should reveal the location of former downward convection currents.

The amount of andesitic volcanic activity probably depends on the rate at which the oceanic plate is being carried down into the mantle for it is this material that, at least in part, feeds the roots of the volcanoes. The interaction of the two plates is also important as the molten rocks will reach the surface more easily where cracks have already been formed. Along the Andes, for example, volcanoes occur where crustal movements produce tensional cracks, but there are no volcanoes where these movements have compressed and closed cracks leading to the surface. With these factors known it should be possible to calculate an average vulcanicity for any given area and so predict the likelihood of volcanic resurgence. This means that the probability of a volcanic eruption occurring can be calculated in active areas and it can be seen if apparently quiescent areas, such as the North Island of New Zealand and parts of the western U.S.A., are likely to remain quiet.

Basaltic volcanoes are fed by mantle rocks from depths of some 50 to 60 kilometres. It was this type of volcanic rock which rose and flooded the land masses through the cracks which opened as the continents separated and is today rising at the oceanic ridges between the plates. Along the Atlantic and India Ocean ridges, these lavas have, in places, built up into high enough peaks to reach the ocean surface where they form the mid-oceanic islands (Figure 5, pages 24–5). In the Pacific, the plates are separating more rapidly so the volcanic peaks along the ridges are much smaller and most basaltic volcanic islands of the Pacific are not associated with the ridge system.

The main economic importance of continental drift is the way in which it controls the distribution of the Earth's surface minerals. The presence of an ore deposit on one side of an ocean obviously suggests its occurrence on the facing continent, but, much more fundamentally, continental drift directly affects the processes by which minerals are concentrated into economically exploitable ores; for example, copper which forms 0·01 per cent of the Earth's crust is found in most rocks but can only be extracted economically from rocks containing at least $1\frac{1}{2}$ per cent.

Concentration takes place in two main ways. Primary ores, such as copper, silver, lead, zinc, manganese, etc., are concentrated by the action of heat and pressure within the Earth's crust and are associated with geological conditions such as granitic intrusions when hot liquids and gases rise towards the surface. These condense as ores at different levels and, as overlying rocks are eroded, become exposed at the surface. Secondary ores are concentrated as a result of weathering and erosion. As rocks are eroded, the non-soluble residue may contain important mineral concentrations such as the bauxite of Arkansas and West Africa. The eroded particles tend to be sorted during transport into their different weights, thereby concentrating, for example, diamonds (South West Africa), tin (Malaya) and gold (Yukon) into placer deposits, while the minerals carried in solution by the streams and rivers may be deposited in basins after evaporation, such as salt (Britain and Germany) and nitrates (Chile), or they may be extracted by plants and animals that concentrate them into coals, limestones, phosphates, etc.

The formation of primary ores obviously depends on crustal rocks being heated and this occurs in two main ways, both controlled by plate movements. When great thicknesses of sediments accumulate, such as occurred in the troughs over the downward convection currents which bordered the Gondwanan and Laurasian continents, then the lower parts of the trough become hotter partly from the heat of the Earth's interior but mainly from the heat given off by the breakdown of radio-active minerals within the sediments. Thus most granites have formed in such sedimentary troughs, and these were mostly developed at or near the sites of downward convection currents. The other source of major heating is at the rising convection currents where the crust is heated directly from the Earth's interior, such as occurred at the cracking apart of our continents. Very large deposits of copper, iron, manganese, gold, silver, and zinc have recently been discovered in hot spots within the Red Sea – an area where continental crust is currently being heated by a rising convection current now separating Arabia from Africa. Similarly, the silver, lead, and zinc deposits of Ireland and Newfoundland formed some 350 million years ago in the fractures which were later to open and form the North Atlantic. These observations suggest that similar rich deposits will be found in certain other parts of the world: in the Labrador Sea between Greenland and Canada which, as we saw in Chapter 8, has been a major zone of weakness, and along the troughs in Africa and northern South America (Figure 33(c), page 93) where the South Atlantic attempted to split through these continents before adopting its present pattern.

The formation of secondary ores is largely climatically controlled and therefore the history of continental movements is extremely important in prospecting for them. Most secondary ores, such as bauxite, oil, and the evaporites, can only accumulate in large quantities in tropical latitudes. Therefore, the knowledge of the location of ancient tropical latitudes (Figure 16, page 50) enables us to explain why, for instance, no oil occurs in East Africa and yet can be found in Europe and northern Alaska; much of North America and Europe have been in tropical latitudes during at least part of the last 400 million years, but East Africa has been further south and only entered tropical latitudes during geologically recent times (Figure 18, page 57 and Figure 33, pages 92–3).

It seems probable that studies of palaeolatitudes may also help us to explain the occurrence of ice caps as these do not appear to be a permanent feature of the Earth's climate. In Chapter 5 we described how the south polar ice cap some 400 million years ago covered the region of the Sahara, then South Africa and Brazil, and by 250 million years ago India, Australia, and Antarctica. The ice cap then disappeared, almost instantaneously, from all of the southern continents and the climate of the world appears to have been less extreme, suggesting that there was no ice cap until some 3 million years ago when ice reappeared at the north pole, although earlier, about 30 million years ago, in the south. It appears that ice caps form either when the continents move to enclose a polar sea or when the continents themselves drift into polar areas and become glaciated as the snow covering them reflects back much of the sun's radiation to reducing the temperature more and more. When there is no land near the polar regions the circulation of oceanic waters from the tropics prevents the accumulation of polar ice and so ameliorates the climate of the whole Earth. This, of course, is rather hypothetical but does offer a cause of major ice ages and suggests that today's ice caps will remain for many millions of years before Antarctica will drift away from the South Pole or the northern continents away from their present circum-polar configuration allowing sufficient tropical waters to enter the Arctic to melt the ice.

In the long term, the economic aspects outlined above will probably be less significant than the way in which this revolutionary view of the Earth is developing and causing Earth scientists to rethink some of the fundamental principles of their subjects and to modify these and face the new problems that arise from this knowledge. Just one example of this is the way we now understand how the present oceans have evolved (Figure 43) but our new knowledge of the oceanic sediments raises the problem of why, between their origin and 70 million years ago, the oceans were apparently shallow, with only weak submarine currents, and then, just as the main opening of the oceans commenced, they deepened and strong oceanic currents began to circulate which have shaped our present continental shelves.

We still know very little about the way in which mountains are formed, but their general evolution is now becoming clearer. We have seen how downward convection currents form deep troughs (like our present oceanic trenches) allowing great thicknesses of sediments to accumulate. These sediments, some 8 to 10 kilometres thick, are lighter than the rocks on either side of the trough and, therefore, rise if they are not held down by the descending convection current, and so a delicate balance develops, particularly when the lower sediments melt from their own radio-active heating to form granite. When the balance changes between the uprising and descending forces, the rocks within the trough become increasingly distorted and when the uprising forces become much greater than the descending forces, which would occur if the convection current either slowed or stopped, there is a major upwelling of the trapped sediments and molten granite. This rise carries the overlying sediments to heights from which they can flow as huge gravity slides, like those seen in the Alps today, and these slides may be further distorted as deeper rocks continue to escape upwards. The building of some of our present mountain chains was further complicated as Africa and India collided with Europe and Asia, crushing the sedimentary troughs between them. A fuller understanding of the relationship between convection currents and mountain building will come from a detailed study of the changes in the rates of ocean spreading. Present studies suggest that there was a brief pause in spreading about 10 million years ago and this time corresponds to the formation of many major structures within our present mountain chains.

Although the fact of continental drift is now established, there is still some evidence, previously used to 'disprove' the theory, which has not yet been explained. Most of this evidence is from past plants and animals whose distribution does not always agree with the continents in the positions indicated by other geological and geophysical evidence. However, now that we know how and when these continental movements occurred, in general terms if not in detail, we are in a position to re-examine much of this evidence. It is probable that in some cases the evidence is based on the apparent distribution of certain fossils found in only a few isolated outcrops of rocks while their true distribution is, as yet, unknown. However, it is probably true that in certain cases, palaeontologists will find their evidence still apparently at variance

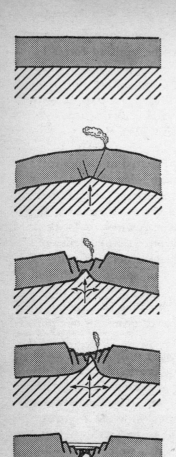

(a) Typical continental crust 30 to 40 kilometres thick overlying typical mantle rocks, e.g. South Africa or Canada.

(b) Arching of the continental mass and widespread volcanic lavas when fractures allow hot mantle rocks to reach the surface, e.g. parts of East Africa.

(c) Release of molton lavas and also slight separation allowing further fracturing and collapse of the arch, e.g. East African Rift Valleys.

(d) Reduction of arching, although vertical movements still predominate. A general depression of the centre allows water to collect and evaporate. Hot springs and volcanic activity persisting with mantle rocks very close to the surface, e.g. the Afar depression, southern Red Sea, today.

(e) Mantle rocks reach the surface and new oceanic crust begins to form as major separation commences. The central areas are now below sea level, but the seas are still very shallow, allowing the accumulation of shallow water deposits of limestone and chert, e.g. the Red Sea today.

43. The evolution of the oceans

(Continued opposite)

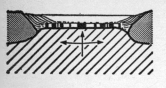

(f) Separation continues with great thickness of sediment accumulating on the continental margins. The oceans, however, are still shallow, so there is little oceanic circulation and, away from the margins, sediments accumulate slowly and contain much organic material, such as limestone. The new oceanic crust being added carries the magnetic pattern of changes of the polarity of the Earth's magnetic field, e.g. the central Atlantic about 120 million years ago.

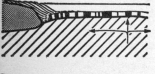

(g) Separation continuing but the oceans are still shallow, e.g. the Atlantic Ocean 70 million years ago.

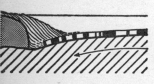

(h) Separation continuing, but the oceans have deepened, initiating major oceanic circulation which moulds the continental shelves into their present shape. Deposits in the deep ocean are variable, often containing material slumping from the continental shelves into the ocean basins, e.g. the Atlantic during the last 60 million years up to the present.

(i) Zones of weakness at the continental–oceanic margin form the sites of downward convection currents and trap sediments in troughs which form the next mountain ranges of the Earth, e.g. the Peru–Chile Trench, Tonga Trench, East Indies, etc.

with that expected from continental drift and it will be interesting to see what other factors, such as wind circulation in the case of plants and predators in the case of animals, affected their distribution on the great, ancient land masses. Therefore, by reversing some of the arguments previously used to demonstrate for or against the reality of continental drift we should reach a fuller understanding of the ancient world. There is, for example, some evidence that the number of species evolved during the last 150 million years has been greater than previously and, although this observation may only reflect the ease of finding evidence from these more recent rocks, it could well be the result of the new geographical isolation of the continents producing a greater number of separate environments. Quite clearly the explanation of the distribution of past climatic belts by moving the whole outer shell of the Earth relative to the Earth's axis of rotation ('true' polar wandering mentioned in Chapter 5) is unnecessary, although it is still possible that a small amount of 'true' polar wandering may occur.

Our present knowledge of continental drift is still very general and much more detailed studies will have to be made to answer the many questions still outstanding, such as – where is the present continental margin of East Africa? Until this is located it is difficult to determine the time when Madagascar separated from Africa and to make a more precise fit of eastern Gondwanaland against Africa. Our knowledge of the Far East and its movements is even more sparse than that for East Africa. How did the Atlantic-Indian-mid-oceanic ridge, which originated around the coastline of Gondwanan Africa, reach its present position? *If* the oceans were all shallow some 70 million years ago, where was all the water that now fills the ocean basins? Why does the magnetic pattern of the ocean floor suggest that the magnetic rocks are all dykes, rising to the surface but not flowing out as lavas over the ocean floor? This may simply reflect the rapid cooling which would occur as soon as these hot rocks contact the water, in which case little extra volcanic water vapour or gases would be released into the oceans and atmosphere, but if lavas do spread out yet do not contribute to the magnetic pattern observed at the ocean surface then a tremendous volume of volcanic gas must have been added to the Earth's atmosphere during the last 100 million years.

Some problems of the convection currents may be solved in the near future when an explanation is found for the linear basaltic islands of the Pacific which began to form well away from the oceanic ridges some 150 million years ago and are still active today. The Hawaiian, Society, Tubai, Tuamoto, and Samoan lines of volcanic islands have each formed from one active centre and have formed separate islands as the movement of the Pacific plate has carried the volcanic piles away from the erupting area.

Perhaps the most fundamental question still remaining is whether the convection currents are shallow and restricted to the 'soft' layer or whether they extend the full depth of the mantle. The answer to this question is particularly important in our understanding of the thermal history of the Earth. We now know that the planets originated from a cold dust cloud in which particles segregated, the denser ones moving nearer the Sun and the lighter ones farther away, probably due to the interaction of magnetic fields in the solar system and the Sun's radiation, both of which were much stronger some 4,500 million years ago. For some reason the dust particles aggregated into clumps of material which grew into our primordial planets. Within these growing planets, heat began to accumulate from the impact of these aggregations, something like meteorites, and also from heat generated by the decay of radio-active materials. In small planets, such as the Moon, this heat was rapidly conducted to the surface and lost into space and only small pockets of molten rocks could form, but the larger planets, such as the Earth, lost their heat much more slowly and the interior of these planets became warm enough for convection currents to develop which segregated the heavier elements downwards to form a core and the lighter elements upwards to form a crust. The Earth, however, could never have become fully molten, as we would then have lost most of the lighter elements, such as hydrogen, into space, but the circulation of these currents collected the lighter 'scum' into a crust over the descending current, probably forming the first primitive continent some 3,500 to 4,000 million years ago. At this point we must reconsider our ideas of the further development of the Earth, for if the convection currents flow throughout the mantle, they would have transported heat to the surface at a much faster rate than has previously been assumed from the conduction alone. However, it does seem as if several crustal nucleii developed which grew by the addition of segments of newly differentiated material. These nucleii, some 3,500 million years old, can now be distinguished in some continents – Canada, Australia, Europe, Siberia – but the story in Africa seems to be more complex as a whole series of ancient discs of continental rock became welded together as recently as 550 to 650 million years ago. If these convection currents still flow throughout the mantle, then the present continental drift is merely one stage in an ever-changing pattern which may, in part, be controlled by the growth of the Earth's core. On the other hand, if the convection currents are shallow, as seems to be most likely, then the changing continental configurations may have no simple pattern.

Meanwhile it is interesting to speculate on the future of our present continents. As we have seen, the oceanic crust is forming faster in the Pacific Ocean than in the other oceans of the world, with new crust spreading out at over 6 cm a year in each direction away from the East Pacific Rise. Nevertheless, the actual area of the Pacific basin is shrinking for it contains nearly all the trenches of the world where the oceanic crust is being carried down into the mantle. This suggests that, at some future date, perhaps another 50 million years, east and west will meet as the Rockies and Andes come into contact with Japan, the Philippines and Tonga. Before then, however, the Mediterranean will disappear as the Alps and Atlas are further compressed to form an even more massive mountain chain through this area. However, interesting as this and many other speculations may be, there remain many fields to study before we fully understand the present situation and can begin to apply this new knowledge of the Earth to the service of Man.

BIBLIOGRAPHY

The books and articles particularly recommended for the general reader are marked with an asterisk. Some of the ideas and information introduced in the later chapters are new and it is suggested that the most recent issues of *Scientific American, Science, Nature, New Scientist*, etc., be consulted for up to date information.

HISTORICAL AND GENERAL GEOLOGY

* A. Holmes, *Principles of Physical Geology* (2nd edn), Nelson, 1965.

A. L. du Toit, *Our Wandering Continents*, Oliver & Boyd, 1937.

A. Wegener, *The Origin of Continents and Oceans* (English translation of the 4th edn of 1929, with new introduction), Methuen & Co., 1967.

L. C. King, *The Morphology of the Earth* (2nd edn), Oliver & Boyd, 1967.

* R. Fraser, *Understanding the Earth*, Penguin Books, 1967.

GEOMETRICAL AND GEOLOGICAL ARGUMENTS

Royal Society, 'A Symposium on Continental Drift', *Proc. Roy. Soc.*, A, 258, 1–323, 1965.

H. Martin, 'The Hypothesis of Continental Drift in the light of recent advances of geological knowledge in Brazil and South-west Africa', *Trans. Geol. So. S. Afr.*, Vol. 64, 1–47, 1961.

* G. A. Doumani and W. E. Long, 'Ancient Life of the Antarctic', *Scientific American*, Vol. 207, 169–84, 1962.

BIOLOGICAL AND CLIMATIC ARGUMENTS

D. V. Ager, *Principles of Paleoecology*, McGraw-Hill, 1963.

* B. Cox, *Prehistoric Animals*, Hamlyn, 1969.

Systematics Association, *Aspects of Tethyan Biogeography*, System. Assoc., Publ. 7, 1967.

A. E. M. Nairn (Ed.), *Descriptive Palaeoclimatology*, Interscience, 1961.

M. Schwarzbach, *Climates of the Past*, van Nostrand, 1963.

A. S. Romer, *Man and the Vertebrates*, Vols. 1 & 2, Penguin Books, 1954.

PALAEOMAGNETIC ARGUMENTS

E. Irving, *Paleomagnetism and its application to geological and geophysical problems*, Wiley, 1964.

D. H. Tarling, *Principles and Applications of Palaeomagnetism*, Chapman & Hall, 1971.

OCEANOGRAPHIC ARGUMENTS

M. N. Hill (Ed.), *The Sea*, Vol. 3, Interscience, 1963.

A. E. Maxwell (Ed.), *The Sea*, Vol. 4(1), Interscience, 1970.

* J. R. Heirtzler, 'Sea Floor Spreading', *Scientific American*, Vol. 219, 66–9, 1968.

H. W. Menard, *Marine Geology of the Pacific*, McGraw-Hill, 1964.

M. J. Keen, *An Introduction to Marine Geology*, Pergamon, 1968.

OTHER WORKS

S. K. Runcorn (Ed.), *Continental Drift*, International Geophysics Series, Vol. 3, Academic Press, 1962.

* H. Takeuchi, S. Uyeda, and H. Kanumori, *Debate about the Earth*, Freemand, Copper & Co., 1967.

R. A. Reyment, 'Ammonite Biostratigraphy, Continental Drift

and Oscillatory Transgressions', *Nature*, Vol. 224, 137–40, 1969.

* B. Kurten, 'Continental Drift and Evolution', *Scientific American*, Vol. 220, 54–64, 1969.

J. A. Jacobs, R. D. Russell, and J. T. Wilson, *Physics and Geology*, McGraw-Hill, 1959.

* P. M. Hurley, 'The Confirmation of Continental Drift', *Scientific American*, Vol. 52, 53–64, 1968.

* B. M. Funnell and A. G. Smith, 'Opening of the Atlantic Ocean', *Nature*, Vol. 219, 1328–33, 1968.

R. A. Phinney (Ed.), *The History of the Earth's Crust*, Princeton University Press, 1968.

M. H. P. Bott, *The Interior of the Earth*, Arnold, 1971.

* I. G. Gass, P. J. Smith, and R. C. L. Wilson, *Understanding the Earth*, Artemis, 1971.

* N. Calder, *Restless Earth*, B.B.C., 1972.

S. K. Runcorn and D. H. Tarling (Editors), *Continental Drift, Sea Floor Spreading and Plate Tectonics: Implications to the Earth Sciences*, Academic Press, 1972.

INDEX

f = figure

More about Penguins and Pelicans

Penguinews, which appears every month, contains details of all the new books issued by Penguins as they are published. From time to time it is supplemented by *Penguins in Print,* which is a complete list of all available books published by Penguins. (There are well over four thousand of these.)

A specimen copy of *Penguinews* will be sent to you free on request. For a year's issues (including the complete lists) please send 30p if you live in the United Kingdom, or 60p if you live elsewhere. Just write to Dept EP, Penguin Books Ltd, Harmondsworth, Middlesex, enclosing a cheque or postal order, and your name will be added to the mailing list.

Note: *Penguinews* and *Penguins in Print* are not available in the U.S.A. or Canada